FROM AL-QAEDA TO THE ISLAMIC STATE (ISIS), JIHADI GROUPS ENGAGE IN CYBER JIHAD:

From 1980s Promotion of Use of 'Electronic Technologies' to Today's Embrace of Social Media to Attract a New Jihadi Generation

Steven Stalinsky and R. Sosnow

MEMRI Books
Washington, D.C.

From Al-Qaeda to the Islamic State (ISIS), Jihadi Groups Engage in Cyber Jihad:
From 1980s Promotion of Use of 'Electronic Technologies' to Today's Embrace of Social Media to Attract a New Jihadi Generation

All rights reserved. Printed in the United States of America. No part of this publication may be reproduced or transmitted in any form or by any means, electronic or mechanical, including photocopy, recording, or any information storage and retrieval system, without permission in writing from the publisher.

©2020 by Steven Stalinsky

Published in the United States of America by MEMRI Books

www.memri.org | cjlab.memri.org

Paperback ISBN: 978-0-9678480-5-1
E-book ISBN: 978-1-7344283-1-5
Library of Congress Control Number: 2020910501

TABLE OF CONTENTS

From Al-Qaeda To The Islamic State, Jihadi Groups Engage In Cyber Jihad: An Introduction By Ambassador R. James Woolsey, Former CIA Director And MEMRI Board Of Advisors Member 8

I. Al-Qaeda's Earliest Cyber Activity 10

II. The Emergence Of The Main Al-Qaeda Websites And Forums – And Their Development Into Providers Of Training For Hacking, Sending Viruses, and Other Forms of Cyber Attacks 13

    Early Jihadi Online Activity: Google Blogs, Yahoo-Hosted Websites 13

    Shumoukh Al-Islam, Al-Fida' and AMEF – Al-Qaeda's Online Open University 13

    The First Cyber Jihad Groups Emerge 14

    On Al-Qaeda-Affiliated Websites And Forums: Cyber Jihad Training, Claims Of Cyber Attacks Against U.S. Airports, Government Agencies Including FBI, CIA, Open Source Center, And The White House, As Well As Military Bases 15

        October 2006: Mujahideen Gather Information on Anchorage International Airport 16

        2006-2007: Jihadi Hackers Focus On U.S. Government: Military, FBI, CIA 16

        2009-14: Hacking And Using Drones 17

        February 2010: Online Jihadis Discuss Cyber Targets in U.S. – Including Government Buildings Like White House, CIA HQ 19

        January 24, 2011: Jihadis Plan Cyber Attack on U.S. Government Computers 20

        June 2011: Shumoukh Al-Islam Writer Calls For Establishing 'Center For Electronic Terrorism' – Which Is Now In "Initial Testing Lab Phase" Prior To Targeting The U.S. 20

        March 2013: Al-Qaeda Electronic Army And Tunisian Cyber Army Claim To Have Hacked Pentagon, State Department Websites 20

        March 2013: Al-Qaeda Claims To Have Hacked U.S. Government's Open Source Center 20

April 2013: Using Android Smartphones To Hijack Airplane 21

May 2013: Ansar Al-Mujahideen Forum Discusses 'Anonymous' Cyber Attack That Paralyzed Guantanamo's Wi-Fi 21

September 2014: ISIS-Supporter Jihadi Media Platform Forum Posts Instructions For Disrupting And Downing Drones Used By U.S. In Iraq 21

Al-Qaeda's Embrace of Encryption Technology 2007-2014 21

III. Statements By Al-Qaeda Leadership On The Importance Of Cyber Jihad 27

Osama bin Laden: "The Wide-Scale Spread Of Jihadist Ideology, Especially On The Internet... [Is] A Major Achievement For Jihad" 27

Ayman Al-Zawahiri On Cyber Jihad: "A Great Front of Islam" 27

Adam Gadahn, American Al-Qaeda Spokesman – And Jihadi Media And Cyber Pioneer 28

Islamic State Of Iraq's Minister Of War: "We Believe Electronic Warfare Is [The] War Of The Future" 28

Senior Al-Qaeda Commander to Potential Recruits: We Need "Specialist Cadres," Not "Regular Fighters" In Afghanistan-Pakistan 28

AQIM Publishes First Installment In Electronic Jihad Series 29

Jihadis Discuss Reported Breach Of TOR, Complain That Tech-Savvy Forum Members Have Abandoned Forums For Twitter 30

AQAP Commander Calls On Followers To Learn Tools for Cyber Jihad From The Internet 32

AQAP Deputy: Online Jihad Is A Great Front And Part Of The Coming Phase 32

Al-Qaeda Deputy Leader and Former Osama bin Laden Secretary Nasir Al-Wuheishi Calls For Volunteers to Join Al-Qaeda Via Encryption Software 33

Anwar Al-Awlaki – The "Bin Laden Of The Internet": The Internet Is A Great Medium For Spreading Jihad 33

Inspire Editor Samir Khan Warns The West That Al-Qaeda Cyber Activists Are Studying Internet Security, Praises The Impact Of Made-In-Afghanistan Jihadi Videos That Are Distributed In The Streets Of London And California 34

U.S.-Designated Global Terrorist Abu Adam Al-Almani: Al-Qaeda's "Professional Media Work" In German And English "Reached Us In Germany" 35

IV. A Major Shift In Online Jihad: From Forums To Social Media 36

Following Killing Of Al-Awlaki and Khan, Al-Qaeda Cyber Activists Promise To Spread Online Jihad And Raid Facebook And Twitter 36

American Al-Qaeda Spokesman Adam Gadahn: "We Must Make Every Effort To Reach Out To Muslims Through New Media Like Facebook and Twitter" 36

Bali Bombing Mastermind: "This Is The Internet Era, There Is Facebook, Twitter, And Others" 37

Taliban Spokesman Praises Impact Of Online Jihad And Use Of Facebook And Twitter 37

Jihadi News Agency "Kavkaz Center," Affiliated With Designated Terrorist Organization "Caucasus Emirate," Calls For Followers To Use Twitter and Facebook 38

Former Guantanamo Detainee And Al-Qaeda Cyber Activist Praises Online Jihad: "Especially Through Twitter And Facebook" 38

V. Western Social Media Companies – At The Heart Of Al-Qaeda, ISIS, Other Groups' Jihadi And Outreach Efforts 39

U.S.-Based Social Media Companies – The Engine Of Jihad Today 39

The San Francisco-Based Internet Archive – Platform For Uploading And Downloading Al-Qaeda Content: Fast, Free, And Unobstructed For Terror Organizations 40

YouTube – The Internet's Primary and Rapidly Expanding Jihadi Base 40

Twitter – Hashtag Jihad And Fundraising For Jihad 41

Friending Al-Qaeda On Facebook 44

Designated Terrorists And Terrorist Organizations Online: Maintaining Official Websites, Using Google Blogspot, Using Yahoo Server, Launching Internet Radio Stations 47

VI. Social Media In The Syria And Iraq Conflict 49

Calls To Engineers To Join The Islamic State July 2014 – By Islamic State (ISIS) Caliph Al-Baghdadi And Others, Including Foreign Jihadis 60

IS Leader Al-Baghdadi – The New "Caliph" – Issues A "Special Call" To Engineers To Join The Islamic State 60

Canadian Jihadi In Syria: Come Fight In Syria – "We Need The Engineers" – And "If You Can't Fight... You Can Assist In Technology" 61

Jihadi Hacktivist Groups Emerge In ISIS 62

Death Photos, Eulogies For Jihadis Killed In Battle Used As A Recruitment Tool 64

Skype – Fundraising And Media Interviews 65

Whatsapp – Mobile Jihadi Messaging 66

Google Services – Mapping, Blogging, And Apps 67

Instagram – Sharing Photos Of Al-Qaeda Leaders – And Used By Jihadis In Syria And Iraq 70

Flickr – Snapshots Of Martyrdom 71

Tumblr – Microblogging Jihad 71

Ask.fm – Jihadi Q&A; Kik – More Privacy 72

SoundCloud – Jihadi Recordings 76

ISIS's Extensive Use Of Social Media 78

Ahrar Al-Sham Leader Hassan Abboud: "It's Very Strange How [ISIS] Has Been Able To Advertise Gory Executions And Beheadings Against The Social Media Websites Rules" 78

ISIS "Media Points" Campaign Provides Religious And ISIS Information To The Public 79

Friendica And Diaspora 83

VK.com 84

JustPaste.it 86

Algorithms: Helping Jihadis Find Each Other – Even After Social Media Accounts
Are Shut Down..86

VII. The Future of Online Jihad – The Coming Battle With The Cyber Army Of Al-Qaeda And Its Offshoots 88

VIII. Endnotes..96

From Al-Qaeda To The Islamic State, Jihadi Groups Engage In Cyber Jihad: An Introduction By Ambassador R. James Woolsey, Former CIA Director And MEMRI Board Of Advisors Member

One of the most important challenges facing the United States and Western world is that of cybersecurity, and understanding the intentions and capabilities of jihadi groups in this realm. It therefore gives me great satisfaction to introduce an historic study that the Middle East Media Research Institute's (MEMRI) Jihad and Terrorism Threat Monitor (JTTM) has been working on for the past year. Given the current situation of Western recruitment to jihad in Iraq and Syria, the information in this study could not be timelier.

ONE CAN HARDLY IMAGINE THE DEVELOPMENT OF THE GLOBAL JIHAD MOVEMENT TO ITS PRESENT PROPORTIONS WITHOUT THE INTERNET – AND AT THE HEART OF THE JIHADI ORGANIZATIONS' STRATEGY ARE U.S. SOCIAL MEDIA COMPANIES.

One can hardly imagine the development of the global jihad movement to its present proportions without the Internet – and at the heart of the jihadi organizations' strategy are U.S. social media companies. Over the past few weeks, senior government officials, including the heads of the FBI and CIA have been discussing the Islamic State's (ISIS) and other jihadi groups' dependence on social media. Last month, FBI director James Comey said that the Islamic State's "widespread use of media and growing online support intensified following the commencement of U.S. airstrikes in Iraq." In addition, the 9/11 Commission Review report noted in July that "cyber attacks can constitute another form of asymmetric terrorism... Security officials are concerned that terrorist groups' skills in computer technology – and in particular in manipulating offensive cyber capabilities – will increase in the years ahead..."

This following report documents jihadi use of the Internet, from Al-Qaeda's and other groups' earliest websites and forums in the 1980s to what we see today with the Islamic State: highly professional video productions and widespread presence on social media, which are integral for recruiting and training the next generation of jihadists.

THE STUDY WILL BE A VITAL CONTRIBUTION TO UNDERSTANDING THIS PHENOMENON, AND EVEN MORE IMPORTANTLY IN DISCUSSING POSSIBLE WAYS OF COUNTERING IT; IT ALSO COULD NOT BE TIMELIER, AS IT IS BEING RELEASED TO COINCIDE WITH THE LAUNCH OF MEMRI'S LATEST INITIATIVE, THE CYBER AND JIHAD LAB (CJL)... IT TOO WILL BE AN IMPORTANT CONTRIBUTION TO THE EFFORT.

The study will be a vital contribution to understanding this phenomenon, and even more importantly in discussing possible ways of countering it; it also could not be timelier, as it is being released to coincide with the launch of MEMRI's latest initiative, the Cyber and Jihad Lab (CJL). This initiative monitors, tracks, translates, and researches cyber jihad originating from the Middle East, Iran, South Asia, and North and West Africa. It translates information from Arabic, Farsi, Urdu, Pashtu, Dari, and other languages into English, produces detailed analyses, and innovates and experiments with potential solutions for stopping cyber jihad. It too will be an important contribution to the effort.

What you are about to read chronicles Al-Qaeda's earliest cyber activity; the emergence of the main Al-Qaeda websites and forums and their development into providers of training for hacking, spreading viruses, and other forms of cyber attacks; statements by Al-Qaeda leadership on the importance of cyber jihad; and the major shift in online jihad from jihadi forums to Western social media, which is now depended upon by jihadis for outreach efforts in the current Syrian and Iraqi conflicts. It explores how nearly a decade has now passed since the U.S. government first pledged to deny terrorists use of the Internet, whereas jihadi activity in cyberspace seems to grow daily. This generation's activists of Al-Qaeda and its offshoots, led by the Islamic State, are younger and Internet savvy, having heeded previous Al-Qaeda leaders' calls to turn to the Internet. They are connected via Facebook, Twitter, YouTube, Instagram, Flickr, and every other emerging social media platform, adopting them almost as soon as they are created – just like the younger generation in the West. Like their Western counterparts, they have smartphones, tablets, and other devices; and, like everyone else, they purchase and use apps.

THIS LANDMARK STUDY SHEDS LIGHT ON A HUGELY IMPORTANT AREA THAT WESTERN GOVERNMENTS, MILITARIES, AND ACADEMICS KNOW TOO LITTLE ABOUT. IT SHOULD BE READ BY EVERYONE IN WASHINGTON – FROM THE DEPARTMENT OF HOMELAND SECURITY AND LEGISLATORS ON CAPITOL HILL TO THE PENTAGON CYBER FORCE, AND THOSE IN ACADEMIA STUDYING THE CYBER REALM.

It is important to note that with the combination of social media and mobile devices, jihadi outlets can make sure that their content is viewable anywhere, anytime. Jihadis were quick to use YouTube and other video sharing services, taking advantage of this technology to provide courses and training in explosives manufacture, weapons training, and hacking. Today, anyone can receive tweets or Facebook posts from Al-Qaeda and its offshoots, led by ISIS, and other terrorist groups, directly to their cellphones, in real time – and can immediately share them far and wide. Another aspect of cyber jihad is hacking; jihadis and jihadi groups are already obtaining funds and wreaking havoc by hacking financial institutions and individuals. This will continue to become more widespread in the future.

In the photos that they now disseminate on social media, the jihadis, who used to pose in their traditional garb surrounded by weapons such as assault rifles and grenades, nowadays include in these photo images a laptop, smartphone, and tablet, reflecting the importance placed on these "weapons." Fittingly, these photos are often chosen as the profile images for their social media accounts.

This landmark study sheds light on a hugely important area that Western governments, militaries, and academics know too little about. It should be read by everyone in Washington – from the Department of Homeland Security and legislators on Capitol Hill to the Pentagon cyber force, and those in academia studying the cyber realm.

Ambassador R. James Woolsey is a former director of the CIA, former under-secretary of the Navy, and former Ambassador to the Negotiation on Conventional Armed Forces in Europe (CFE). He is also a member of MEMRI's Board of Advisors.

I. Al-Qaeda's Earliest Cyber Activity

Over the past 20 years, Al-Qaeda and other jihadi groups have been quietly investing in their cyber jihad capabilities. As Abu Ubayd Al-Qurashi, one of Osama bin Laden's closest aides, warned the U.S. in 2002, "jihad on the Internet" has now become one of the "nightmares" to be faced in the future.[1] Al-Qaeda news and information, once conveyed to followers via messengers, fax, and the Qatari television network Al-Jazeera, is now promoted by a new, Internet-savvy generation of jihadis. From their very earliest ventures into this sphere that included jihadi websites, forums, and discussion groups, whose members were carefully screened, they have moved onto today's social media platforms. They use these platforms not only to spread their messages, but, as this report will show, to recruit activists with computer and Internet skills who are actively involved in studying how to hack the websites of government institutions and banks, hijack drones and aircraft, and carry out other cyber crimes.

With the latest transition by Al-Qaeda and its offshoots, including ISIS, to social media, all of this content is now accessible to the entire world. Anyone, from journalists to the general public, can now follow, "like," "friend," re-tweet, submit questions to, view photos posted by, and conduct dialogue with Al-Qaeda and other jihadi groups online – and can just as easily be recruited by it. The advent of social media has changed the status quo; while Al-Qaeda and the main jihadi media groups have lost a great deal of control over their own public relations efforts, they are now reaching a much larger – and constantly growing – pool of potential recruits.

The previous main jihadi forums were always private, with their own internal hierarchy; there was an online application process for anyone requesting to join and they also had to be vouched for by trusted forum members. Thus, the circle of online jihadis was limited. From these closed forums, the online jihad of Al-Qaeda and its offshoots expanded to countless websites and other, less important forums and blogs that were open and that primarily reposted and redistributed content. Jihadi media groups were established, and the new content that they provided included increasingly professional video productions and online magazines, including in English as well as in European languages.

Since the launch two decades ago of what is considered one of the "first real Al-Qaeda website[s],"[2] Azzam.com, whose operators pleaded guilty to terrorism charges in December 2013 in a U.S. federal court, the importance of cyber jihad has been discussed by the Al-Qaeda leadership. This includes Osama bin Laden and Ayman Al-Zawahiri as well as the heads of Al-Qaeda's offshoots Al-Qaeda in the Arabian Peninsula (AQAP), Islamic State of Iraq (ISI), the Somali Al-Qaeda affiliate Al-Shabaab

Al-Mujahideen, and Al-Qaeda in the Islamic Maghreb (AQIM), as well as by the leadership of other jihadi groups such as the Taliban, and, now, the Islamic State (previously known as the Islamic State in Iraq and Syria, or ISIS) and others.

'Abd Al-Bari 'Atwan, the pro-jihad author and former editor of the London daily *Al-Quds Al-Arabi*, who in 1996 interviewed Osama bin Laden and is known to be close to Al-Qaeda, wrote in a chapter titled "Cyber Jihad" in his 2008 book *The Secret History of Al-Qaeda* that by the mid-1980s, Sheikh Abdullah Azzam, the Palestinian Sunni Islamist cleric and Osama bin Laden mentor who convinced bin Laden to come to Afghanistan to aid the jihad, was "encouraging jihadi groups to exploit the potential of evolving electronic technologies." 'Atwan added: "The Internet has become a key element in Al-Qaeda training, planning, and logistics, and cyberspace a legitimate field of battle. Some commentators have gone so far as to declare that Al-Qaeda is the first Web-directed guerrilla network."[3]

Pakistani journalist and bin Laden biographer Hamid Mir, who in the summer of 2014 was attacked by gunmen in Karachi, noted, as he watched Al-Qaeda members fleeing a U.S. bombardment of their training camps in November 2001: "Every second Al-Qaeda member [was] carrying a laptop computer along with his Kalashnikov."[4]

One early Al-Qaeda recruit, the Moroccan L'Houissaine Kherchtou, was sent by the Al-Qaeda leadership to learn high-tech methods of surveillance from Abu Mohamed Al-Amriki ("The American"), the Egyptian-born American who was a key aide to bin Laden.[5] Kherchtou and other trainees learned about potential targets such as bridges, major sports stadiums, police stations, and consulates. He then joined Al-Qaeda's electronic workshop at Hyatabad in Peshawar, Pakistan, according to his February 21, 2001 testimony in United States of America v. Usama bin Laden, et al.[6]

Another early figure in Al-Qaeda cyber activity was Irhabi 007, aka Younes Tsouli, a 23-year-old Moroccan-born resident of the U.K. In 2007, he was found guilty of incitement to commit acts of terrorism, and sentenced to 16 years in a U.K. prison. Tsouli, considered the godfather of online jihadis, was contacted by Al-Qaeda in Iraq leader Abu Mus'ab Al-Zarqawi and promoted and spread Al-Qaeda online.[7]

The "sacred duty" of cyber jihad was codified by Al-Qaeda in a 2003 Al-Qaeda document called "The 39 Principles of Jihad," on the Al-Farouq website, a known Saudi-based Al-Qaeda domain. This duty includes participation in online forums to defend Islam. "Al-Salem," who posted the document, noted that the Internet offers a way of reaching millions of people in seconds, and that those with Internet skills are encouraged to use them to support the jihad by hacking "enemy" websites as well as "morally corrupt" ones such as pornographic sites.[8] Years later, Bostonians Ahmad Abousamra, who is on the FBI's Most Wanted Terrorists list and is thought to be currently in Syria helping ISIS,[9] and Tarek Mehanna, who is serving 17 years in federal prison after being found guilty of translating material for Al-Qaeda in Iraq, were found to have translated the "39 Principles" into English and posted it on a list serve called At-Tibyan Publications, where English-speaking Muslims praised Al-Qaeda.[10]

Defense, intelligence, security, and military authorities are becoming increasingly aware of the issue of terrorists' use of the internet as part of cyber security. For example, at a January 29, 2014 U.S. Senate Select Committee on Intelligence hearing, Defense Intelligence Agency director Lt.-Gen. Michael Flynn added: "I think that what is a serious threat that we are paying very close attention to are these non-nation-state groups and actors, Al-Qaeda being among them... that are also operating in the cyber domain... [T]hey are increasingly adapting to an environment that is actually benefiting them." He added that this is "an increasingly growing threat." But, as this report shows, such online activity has been developing over the past two decades.

John Carlin, the Assistant Attorney General for National Security in the U.S. Department of Justice, said on July 24, 2014 at the Aspen Security Forum session titled "U.S. Counterterrorism Strategy and its Implications for Emerging National Security Threats": "[L]ooking ahead on cyber terrorism – and the 9/11 report just gave an updated report – we're in a ... pre-9/11 moment with cyber. It's clear that the terrorists want to use cyber-enabled means to cause the maximum amount of destruction as they can to our infrastructure. It's clear because they have said it. [Al-Qaeda leader] Zawahiri put out a videotape statement saying, 'We want to commit a cyber attack against the United States.' It's also clear that other actors – nation-state actors... have the capability now to cause significant damage..." The 9/11 Commission's report to which Carlin was referring, released July 22, 2014 and titled "Reflections on the Tenth Anniversary of *The 9/11 Commission Report*," warned: "Cyber attacks can constitute another form of asymmetric terrorism... Security officials are concerned that terrorist groups' skills in computer technology – and in particular in manipulating offensive cyber capabilities – will increase in the years ahead..."

At the heart of jihadi operations today in Syria and Iraq, and other places, are Western social media companies; cyber space is now part of their arsenal and will only become more central to them in the future. As Robert Hannigan, newly appointed director of Britain's Government Communications Headquarters (GCHQ), which works closely with MI5, MI6, and the U.S.'s NSA, wrote in the *Financial Times* on November 3, 2014, said of "the largest U.S. technology companies which dominate the web": "However much they may dislike it, they have become the command-and-control networks of choice for terrorists and criminals, who find their services as transformational as the rest of us. If they are to meet this challenge, it means coming up with better arrangements for facilitating lawful investigation by security and law enforcement agencies than we have now... [W] we need a new deal between democratic governments and the technology companies in the area of protecting our citizens."

To highlight how important the cyber realm is to them, jihadis, who in previous years commonly posed for photos wearing their traditional garb and surrounded by weapons such as assault rifles and grenades, today pose similarly but with a laptop, smartphone, and tablet added to their arsenal; they frequently choose these images as profile images for their social media accounts. This report will summarize the development of Al-Qaeda's and its affiliate groups' cyber efforts, from the earliest closed online jihad websites and forums to its transition into social media accessible to all.

II. The Emergence Of The Main Al-Qaeda Websites And Forums – And Their Development Into Providers Of Training For Hacking, Sending Viruses, and Other Forms of Cyber Attacks

On December 10, 2013, Babar Ahmad and Syed Talha Ahsan pleaded guilty in a U.S. federal court to providing material support to terrorists. The charges stem from their involvement in and operation of the London-based Azzam Publications and its associated website Azzam.com, which went online in 1994 and which provided material support to the Chechen mujahideen, the Taliban, and associated terrorist groups.[11] According to *The Washington Post,* Azzam.com was "the first real Al-Qaeda website" which "taught an entire generation about jihad."[12] Azzam.com also recruited fighters and raised funds. It was reported in 2002 that the website was receiving five million hits a day from across the world.[13]

Another early Al-Qaeda website was Maalemaljihad.com ("Milestones of Jihad"),[14] established in 2000 and hosted in China with a mirror site established in Pakistan a few months later. The content, including statements by bin Laden and Al-Zawahiri, and the Al-Mujahidoun newsletter, was provided from Afghanistan. The site crashed a year later, and the Pakistani mirror site crashed the summer after that.[15]

The MEMRI Jihad and Terrorism Threat Monitor (JTTM) has been monitoring the development of Al-Qaeda's and other jihadi groups' online presence for over a decade. The most important forums began with Al-Neda in 2001, and expanded to Al-Hesbah, Al-Ekhlas, and Al-Falluja, all now defunct. Currently, there is only a handful of main Arabic-language jihadi forums. They are Ansar Al-Mujahideen, established in 2008 in Arabic which, while currently inactive, was active as late as the spring of 2014, and which later launched multiple language pages, including English;[16] Shumoukh Al-Islam, which has been in existence since at least 2009; Al-Fida', which was launched in 2011;[17] and, most recently, the pro-ISIS forum Jihadi Media Platform (alplatformmedia.com). Over the past year, the forums have become slow and unreliable, and there is speculation that this is because they are under intermittent attack.

Early Jihadi Online Activity: Google Blogs, Yahoo-Hosted Websites

A MEMRI study from July 2007, *The Enemy Within: Where Are the Islamist/Jihadist Websites Hosted, and What Can Be Done About It?*, and other research conducted by MEMRI,[18] found that beginning shortly after 9/11,[19] the Al-Qaeda leadership and other jihadis depended on blogs and websites hosted by Google, Yahoo! and MSN, for two main purposes: operational needs and indoctrination and da'wa (propagation of Islam). From there, they developed into password-protected websites and forums, containing chat rooms and message boards, which they used, and continue to use, as a direct channel for disseminating communiques, online training, and audio and video recordings.[20] These websites and forums, whose members were carefully vetted, included Al-Faluja, Shabakat Ansar Al-Mujahideen, Shabakat Al-Tahadi Al-Islamiyya, Shumoukh Al-Islam, Shabakat Hanein, Shabakat Al-Mujahideen Al-Electroniyya, and Mutadayat Al-Boraq Al-Islamiyya (network).[21]

While, as mentioned, several forums – Shumoukh Al-Islam, in addition to Al-Fida' and AMEF[22] – are still in use today, a great deal of Al-Qaeda's and jihadis' online activity has shifted to largely U.S.-based social media. Currently, there is an internal debate among different groups in the jihadi world about the use of forums versus social media and what future direction online jihad should take.

Shumoukh Al-Islam, Al-Fida' and AMEF – Al-Qaeda's Online Open University

Both Shumoukh and Al-Fida' have sections on their forums devoted to military and technology topics – Shumoukh's is "The Shumoukh [Military] Camp" and Al-Fida's is "Modern Technical Science" – and both provide tools for jihadis to operate online.

Underlining Al-Qaeda's interest in and focus on military and technology is the wide range of instructions available on these forums for making explosives, lessons on attacking targets, and instructions for using online tools, such as hacking, encryption, video production, and more.

Another forum devoting substantial space to providing its readers with access to tools for cyber jihad is the Ansar Al-Mujahideen English Forum (AMEF) which included assorted information such as reports on U.S. government online surveillance and on computer viruses, and instructions for making weapons; links to information related to U.S. government actions against online jihad and tips for circumventing them; and tutorials on disseminating content online, with answers to questions. In fact, the forum even offers a thread for anyone to submit IT questions.[23]

According to a groundbreaking 2007 MEMRI Jihad and Terrorism Threat Monitor (JTTM) report by Dr. Eli Alshech, *Cyberspace as a Combat Zone: The Phenomenon of Electronic Jihad*,[24] not only did these websites and forums provide opportunities for groups to coordinate cyber attacks;[25] they also clearly showed that these attacks were not random initiatives by individual mujahideen but were well organized and overseen by a network of specially appointed individuals on various sites.

The report stated that many of these organized and coordinated attacks were organized by groups devoted to cyber jihad, and that they involved efforts by many forum members. Announcements of imminent attacks, would appear almost daily, posted on numerous Islamist websites simultaneously, with information on how to participate – including the time, the target, and the software to be used for the attack.[26]

The First Cyber Jihad Groups Emerge

Six prominent groups devoted to cyber jihad that emerged in the mid-2000s were Ansar Al-Jihad Lil-Jihad Al-Electroni, Munazamat Fursan Al-Jihad Al-Electroni, Majmu'at Al-Jihad Al-Electroni, Majma' Al-Haker Al-Muslim, and Inhiyar Al-Dolar, and Hackboy; all of these were active as of February 2007, when MEMRI reported on them.[27]

All these groups, with the exception of Munazamat Fursan Al-Jihad and Inhiyar al-Dolar, also had their own websites or forums, through which they recruited volunteers to help carry out the cyber attacks, and for other purposes. For example, Majmu'at Al-Jihad Al-Electroni had several sections, including one devoted to electronic jihad strategy, a technical section about software to use for cyber attacks, a log of previous attacks, and a section including various appeals to Muslims, mujahideen, and hackers worldwide.[28] Messages in these forums were on a number of topics, including eliminating websites that they claimed harmed Islam, causing economic damage to the West and waging psychological warfare against it, and bringing about the total collapse of the West.

A message posted on one of the forums on the topic of eliminating websites that harm Islam stated: "We are indeed victorious when we disable such [harmful] websites, but the matter is not so simple. We target... websites that wage intensive war [against us]... We target them because they are the foremost enemies of jihad in cyberspace; their existence threatens Islamic and religious websites throughout the Internet..."[29] A message on the topic of economic damage and psychological warfare against the West read: "Allah has commanded us in various Koranic verses to wage war against the unbelievers... Electronic jihad utilizes methods and means which inflict great material damage on the enemy and [which also] lower his morale and his spirits via the Internet. The methods of [hacking] have been revealed [to us] by expert [hackers] on the Internet and networks... many of whom engage in purposeless and meaningless sabotage. These lethal methods will be harnessed [for use] against our enemies, so as to inflict the greatest [possible] financial damage [upon them] which can amount to millions – and [in order] to damage [their] morale, so that [they] will be afraid of the Muslims wherever they go and even when they are surfing the Web."[30]

Likewise, a message on the topic of bringing about the total collapse of the West stated: "I have examined most of the material [available] in hacking manuals but have not found articles which discuss... how to disable all the [electronic] networks around the world. I found various articles which discuss how to attack websites, emails, servers, etc., but I have not read anything about harming or blocking the networks around the world, even though this is one of the most important topics for a hacker and for anyone who engages in electronic jihad. Such [an attack] will cripple the West completely. I am not talking about attacking websites or [even] the Internet [as a whole], but [about attacking] all the [computer] networks around the world including military networks, and [networks] which control radars, missiles and communications around the world... If all these networks stop [functioning even] for a single day... it will bring about the total collapse of the West... while affecting our interests only slightly. The collapse of the West will bring about the breakdown of world economy and of the stock markets, which depend on [electronic] communication [for] their activities, [e.g.] transfers of assets and shares. [Such an attack] will cause the capitalist West to collapse."

The Muslim Hackers Group, which was founded in 1998,[31] stated on its website: "It's time for Muslims on the Web, knowledge able of hacking, virus making, and all those fringe matters, to join a Club and share their knowledge."[32] The website reportedly featured links to U.S. sites that claim to disclose sensitive information like code names and radio frequencies used by the U.S. Secret Service, and offers tutorials in viruses and hacking, as well as links to other jihadi web addresses.[33]

In 2002, it was reported that among the recent targets that Al-Qaeda and other terrorists had been discussing online, according to people with knowledge of intelligence briefings, included the Centers for Disease Control and Prevention in Atlanta; FedWire, the money-movement clearing system maintained by the Federal Reserve Board, and facilities controlling the flow of information over the Internet.[34]

On Al-Qaeda-Affiliated Websites And Forums: Cyber Jihad Training, Claims Of Cyber Attacks Against U.S. Airports, Government Agencies Including FBI, CIA, Open Source Center, And The White House, As Well As Military Bases

Al-Qaeda and its offshoots, and their online army of supporters, are testing – and finding – ways to launch cyber jihad, and are planning for future attacks, by hacking, sending viruses, and other methods. They are also claiming to have carried out cyber attacks. This was highlighted by an important Al-Qaeda release in June 3, 2011; in a two-part video by Al-Qaeda's Al-Sahab media outlet titled "Do Not Rely on Others, Take [the Task] Upon Yourself," Part One focuses, inter alia, on cyber warfare. The narrator sets out the conditions for anyone considering taking part, saying that anyone from anywhere with talent in such a field (i.e. computer science) is encouraged to play a role, provided that these participants' actions are "in harmony with the mujahedeen's general plan." He states: "Internet hacking is one of the important avenues of jihad, and we [Al-Qaeda] advise Muslims who possess expertise in this field to target the websites and electronic networks of major corporations and government administrations in the countries [involved in] attacking Muslims... and to focus on websites and networks of media centers that fights Islam, jihad and the mujahedeen."[35]

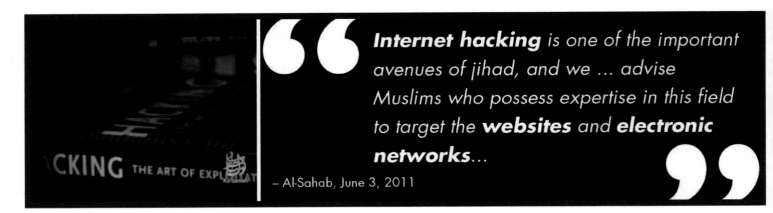

— Al-Sahab, June 3, 2011

The video was taken very seriously by then-Senate Committee on Homeland Security and Government Affairs Chairman Joseph Lieberman and Ranking Member Susan Collins, who called on Congress to pass cyber security legislation because of possible threats by Al-Qaeda.

In addition to those mentioned above, there have been countless more claims of cyber attacks and discussion of potential targets on Al-Qaeda-affiliated websites and forums and other venues in recent years. While it is generally not known whether these attacks were successes or failure – or even if they were attempted at all – such online discussions are constantly ongoing and must be taken seriously. The following examples highlight some of these claims and discussions.

October 2006: Mujahideen Gather Information on Anchorage International Airport

On October 17, 2006, a jihadi website posted a message titled "You Can Spy on the Enemies' Airports Directly by Controlling the Cameras' Direction." The message contains a link to a screen showing what it claimed to be a live view of various areas within Anchorage International Airport via several cameras. The message gives directions for how to control the cameras and promises to provide links in the future for other airports as well.[36]

2006-2007: Jihadi Hackers Focus On U.S. Government: Military, FBI, CIA

In September 2006, the jihadi website Alnusra.net posted a long list of IP addresses to be targeted, claiming that they were associated with key governmental defense institutions in the West, including "the Army Ballistics Research Laboratory," "the Army Armament Research Development and Engineering Center," "the Navy Computers and Telecommunications Station," "the National Space Development Agency of Japan," and more.[37]

A few months later, in December 2006, another jihadi website, Alfirdaws.org, stated that Islamist hackers had canceled a planned attack against U.S. banks, named The Electronic Guantanamo Raid, because the banks had been alerted by U.S. media reports and government agencies. The website claimed that the media panic in the matter showed the importance of "focus[ing] on attacking sensitive economic American websites [instead of] other [websites, like those that offend Islam]." It added, "If [we] attack websites associated with the stock [market] and with banks, disabling them for a few days or even for a few hours, it will cause millions of dollars' worth of damage... I [therefore] call upon all members [of this forum] to focus on these websites and to urge all Muslims who are able to participate in this [type of] Islamic Intifada to attack websites associated with the American stock[market] and banks..."[38]

Additionally, a 2007 Islamist forum's survey of its members about what targets they would like to attack found that Western financial websites and websites associated with the FBI and CIA topped the list.[39]

2009-14: Hacking And Using Drones

Since the U.S. began implementing a strategy of killing Al-Qaeda leaders and members with drones, Al-Qaeda has, in response, invested major efforts in attempts to hack into drones and to stop them by other technological means. Today, the use of drones is of strategic cyber importance not only to Al-Qaeda but to ISIS and other jihad organizations in Syria, Iraq, Yemen, Somalia, North Africa, and elsewhere.

In July 2010, the Defense Intelligence Agency, a U.S. spy agency, intercepted electronic communications indicating that senior Al-Qaeda leaders had distributed a "strategy guide" to operatives around the world advising them how "to anticipate and defeat" unmanned aircraft. The agency reported that Al-Qaeda was sponsoring simultaneous research projects to develop jammers to interfere with GPS signals and infrared tags that drone operators rely on to pinpoint missile targets. According to the report, other projects in the works included the development of small radio-controlled aircraft, or hobby planes, which insurgents apparently saw as having potential for monitoring the flight patterns of U.S. drones. DIA analysts also noted that they believed that Al-Qaeda "cell leadership is tracking the progress of each project and can redirect components from one project to another." That same year, the CIA noted in a report that Al-Qaeda was placing special emphasis on the recruitment of technicians and that "the skills most in demand" included expertise in drones and missile technology.[40]

The cover story of the fifth issue of the Taliban's English-language magazine *Azan,* released March 28, 2014, was titled "Counter-Drone Strategy" and authored by known jihadi writer Jaffer Hussain. In it, he wrote: "We call on our skillful Muslim brothers who are engineers and scientists to come forward and try their best in figuring out how to break the link between drone and GPS. Experiment in whichever part of the world you live and if it's successful, make a complete video demonstration of the process, upload it on the web and make it password protected. Then send that link and password to us. Or simply make a good... PowerPoint [presenta-

"Counter-Drone Strategy," Azan, Issue V, released March 28, 2014.

tion] and send it to us. Even if you have made good progress in the experiment but encountered some complication in it then send it to us, maybe we can suggest something useful to you."

Hussain added, "Our brothers who are computer engineers and programmers – who can hack into the private encrypted network of the Pentagon, try your best to do so. ... If only one brother takes the initiative and attempts to hack the Pentagon, he'll set an excellent example for other brothers to follow suit..." The security check[s] in Muslim countries of Middle East, Subcontinent [of India and Pakistan], and Africa is relatively low as compared to the Western kafir countries. So for example, if you are a university student in Kuala Lumpur (Malaysia) you can download *Azan* magazine or *Inspire* magazine from the Internet, print it, and distribute it safely in your university or masjid [mosque]. Make a fake Facebook account and create a page for *Azan* magazine... Try to find that hacker in Jakarta or Jeddah who can work only for Allah and be the candidate for the luxurious Jannah [Paradise]... Do whatever you can to spread the word of tawheed [monotheism] and jihad! ... [T]his is the phase of the battle in which every Muslim can contribute to the global jihad in whatever capacity they can."[41]

Drones are also of great interest to jihadis in Iraq and Syria and to those who follow their exploits on social media. For example, on March 20, 2014, ISIS used a UAV to capture aerial photos of one of its parades in Fallujah, Iraq, and then tweeted photos of the UAV. The photos – and ISIS's drone capability – were circulated, discussed, and praised by jihadis online.

Twitter.com/ShamiWitness/status/446886806827433984; Twitter.com/dawlat_islam1/status/446887861946957824; Twitter.com/seifaldolh/status/503216673361367040

In 2009, it was reported that Iran-backed Iraqi militants had hacked into video feeds from Predator and Shadow drones.[42] They passed what they found on to Hizbullah.[43]

On October 27, 2014, a member on the jihadi forum Shumoukh Al-Islam recommended using drones to carry out terror attacks in the U.S. and other Western countries. In his post he included a link to a CNN video in Arabic about Amazon's plan to deliver packages via drone. He suggested that Amazon's plan could be adapted for sending parcel bombs aboard UAVs to sensitive targets in the U.S. and the West, and assessed that a lone wolf could securely and anonymously carry out five or six terror attacks in this manner.[44]

ISIS released a video that same day, showing its use of drones in the Iraqi city of Kobane, which included images of Kobane from the air and a grain silo across the border in Turkey.[45]

On October 23, 2014, Jabhat Al-Nusra released a video titled "Breaking the Siege" that featured a commander on his computer as well as images captured from the group's drones:

February 2010: Online Jihadis Discuss Cyber Targets in U.S. – Including Government Buildings Like White House, CIA HQ

A discussion thread begun February 2, 2010 on the Al-Falluja jihadi forum discussed which websites in the U.S. would make the most effective targets for a terror attack – or, as the initiator of the thread put it, "how we can hasten getting rid of America, Allah willing." A second participant in the thread, "Rahiq," wrote that to destroy the U.S. it is necessary to 1) strike at power stations and refineries, and 2) take out communications networks. He recommended that this be coordinated with a cyber attack impacting political and military lines of command, raising the possibility that weapons could then be remotely launched either at U.S. targets or at foreign countries like China or Russia in an attempt to draw them into war with America. He also noted that a sufficient quantity of explosives might trigger a manmade earthquake. A number of other forum members praised Rahiq as a "true terrorist."[46]

January 24, 2011: Jihadis Plan Cyber Attack on U.S. Government Computers

According to a January 24, 2011 post on the jihadi forum Shumoukh Al-Islam by a new member calling himself Osama Al-Amriki (Osama the American), jihadists were planning a coordinated DDOS [distributed denial-of-service] attack against a U.S. government computer system. The attack, which the post said was set for February 5, 2011 at 11:00 GMT, was to take place over a span of four hours. The jihadis' plan was to send thousands of messages to a government website using special programs that cause the system to overload and ultimately crash. The planners said that they would post the specific web address of the facility to be targeted one day before the attack was to begin. Attackers were advised to send the messages from Internet cafés, and to conceal their IP addresses using proxy software.[47]

June 2011: Shumoukh Al-Islam Writer Calls For Establishing 'Center For Electronic Terrorism' – Which Is Now In "Initial Testing Lab Phase" Prior To Targeting The U.S.

On June 11, 2011, Yaman Mukhaddab, a prominent writer on the major jihadi forum Shumoukh Al-Islam, posted a call on the forum to establish a Center for Electronic Terrorism (CET), to specialize in virtual attacks against the U.S., U.K. and France. His post discussed the initial phase of the center, which would include background information about the proposed center and a call for prospective recruits. He presented the CET's priorities and goals as focused on electronically targeting infrastructure in the U.S., U.K., and France.[48]

Fast forward two years to the April 2013 "#opIsrael" coordinated cyber attacks, which was aimed at causing major widespread damage to Israeli websites. Mukhaddab announced that an "electronic jihad brigades project" established two years ago – likely the CET – had participated in the attacks, and referred to #opIsrael as this project's "training field and initial testing lab." He noted that the U.S. is in fact the main intended target for the e-jihadi brigades' activities, and, addressing President Obama, said that the targeting of Israel was only a fraction of the planned "final production" against the U.S. The U.S., he said, presents many "soft bellies" for targeting, which, once targeted, would have a severe damaging impact on the country.[49]

Mukhaddab's revelations about the cyber jihad efforts by Al-Qaeda and its affiliates – not yet targeting the U.S. but in the "initial testing lab" phase – indicate that he is following the directives of the Al-Qaeda leadership. As a prominent writer on the Shumoukh Al-Islam forum, he is uniquely placed as an Al-Qaeda insider who knows the leadership's strategy.

March 2013: Al-Qaeda Electronic Army And Tunisian Cyber Army Claim To Have Hacked Pentagon, State Department Websites

On March 7, 2013, Tunisian Cyber Army (TCA) posted on Twitter[50] its plans to launch a cyber-campaign titled "#opBlackSummer," which it said would extend from May 31 until September 11, 2013. TCA warned that it, along with "Al-Qaeda Electronic Army," it would aim "painful blows" at the U.S. until it achieved "victory" against it. A few days later, on March 10, TCA claimed in an Arabic-language Tweet, "Thanks to Allah, the Al-Qaeda digital [team], in collaboration with us [i.e. TCA], has succeeded in striking America [with] a painful blow against the Pentagon." On the same day, ehackingnews.com reported that a subdomain of the Pentagon purportedly had been hacked by TCA and Al-Qaeda Electronic Army as part of the same campaign. It thus appears that TCA has already begun its cyber attack campaign – sooner than its previously announced date of May 31. Interestingly, the ehackingnews.com report noted that the attack against the Pentagon site was conducted in collaboration with Chinese hackers.[51]

March 2013: Al-Qaeda Claims To Have Hacked U.S. Government's Open Source Center

In another incident, on March 30, 2013, 'Abd Al-Hamid Al-Irlandi ("The Irish"), a Shumoukh Al-Islam forum member who posts regularly, posted in the forum his musings on whether Al-Qaeda was behind an alleged cyber attack that targeted the

U.S. government website Open Source Center (OSC). He did not, however, specify when such alleged cyber attacks might have taken place, nor did he provide evidence of such an attack. "Was the Al-Qaeda organization able to penetrate one of the most important offices of the Open Source Center in Qatar, which specializes in counter-'terrorism' matters?" he wrote, adding, "And was this penetration technical [i.e. via cyber attack] or was Al-Qaeda able to recruit [another] Abu Dujana Al-Khurasani2?"[52] He also speculated that with this alleged cyber attack Al-Qaeda had been able to obtain sensitive information from the personal account of a "senior federal employee."[53]

April 2013: Using Android Smartphones To Hijack Airplane

On the jihadi forum Shumoukh Al-Islam, one writer discussed the possibility of using an Android smartphone to override an airplane's monitoring systems in order to take control of it and hijack it. The writer stated that at an April 2013 security summit in Amsterdam, a German security expert named Hugo Tesco had reported developing a system called Simon that does just that, and that this system contains malicious code that could attack and infiltrate the security system of an airplane using an Android mobile phone application. The writer added that Tesco also said that the plane's speed and flight path could be changed by sending radio waves to its flight management system.[54]

Also on Shumoukh Al-Islam, in late December 2013, one Abu Mujahid Al-Emarati announced the launch of a course in hacking Android-based devices. Al-Emarati also provided links to his YouTube channel, where he has posted several Android hacking video tutorials in Arabic are found. Several of these videos include a disclaimer written in English, which states that the content of the material is for educational purposes only and that it should not be used against Muslims.[55]

May 2013: Ansar Al-Mujahideen Forum Discusses 'Anonymous' Cyber Attack That Paralyzed Guantanamo's Wi-Fi

During the latter half of May 2013, visitors to the jihadist Web forum Ansar Al-Mujahideen discussed a cyber attack by Anonymous against Guantanamo Bay prison. The report added that a prison spokesman had said that a cyber attack on May 20, 2013 had paralyzed the prison's wireless Internet system, and that Anonymous had clarified that the attack had been meant to express support for prisoners who were on hunger strike.[56]

September 2014: ISIS-Supporter Jihadi Media Platform Forum Posts Instructions For Disrupting And Downing Drones Used By U.S. In Iraq

A document posted August 30, 2014 on the Jihadi Media Platform forum (alplatformmedia.com), which supports ISIS, contains instructions for disrupting and downing the drones being used by the U.S. in Iraq. The document, originally posted on the Twitter account @revbaghdad3, was posted in response to the renewed U.S. bombings of ISIS targets and the provision of air support to Kurdish forces combating the organization's advance. In a different post on the same forum, an ISIS supporter suggested making military use of a small unmanned octocopter (eight-rotor helicopter) that is being developed for Amazon to ferry mail parcels.[57]

Al-Qaeda's Embrace of Encryption Technology 2007-2014

The MEMRI JTTM's research has tracked Al-Qaeda's encryption development efforts from basic software first used by a few high-ranking members to mass online distribution available to major Al-Qaeda-affiliated websites and chat rooms.[58] In January 2007, Al-Qaeda began to use encryption tools for its online activities, particularly for communications efforts, often utilizing security software based on military grade technology. The goal was to hide messages and to protect data transferred via networks, the Internet, mobile phones, ecommerce, Bluetooth, and the like. This development was in direct response to various security breaches of its websites over the past years by Western government agencies.

The issue of encryption by Al-Qaeda made news following the killing of Osama bin Laden. U.S. intelligence sources reported that much of the material seized at bin Laden's compound was encrypted and stored electronically on computers, laptops, hard drives, and storage devices.[59] Encryption software created by the Global Islamic Media Front (GIMF) in 2007[60] is relied on by many jihadi groups, and it has been upgraded a number of times. In its most recent upgrade, on July 12, 2014, the Al-Qaeda-affiliated Global Islamic Media Front (GIMF) released an updated version of its Android mobile encryption software, citing as a reason for doing so the fact that "global [communications] companies" are now cooperating with "international intelligence agencies." GIMF released the first version of this software in September 2013.[61]

The newer version, explains GIMF, "enables the encryption of files directly from the mobile Android [device]." The announcement, in English, adds: "With the increased spread and circulation of smartphones worldwide, especially phones running the Android operating system, users in the Arab and Muslim countries have become very reliant on them in their daily life. Smartphones are used for internet access, emails, and navigating news and social networking websites, and thousands are even using smartphones to follow the news of jihad in Islamic countries. This requires a degree of security precaution, in accordance with the words of Allah the Almighty [Koran (4:71)]: 'Take your precautions,' especially in the midst of the rapidly developing news about the cooperation of global companies with the international intelligence agencies, in the detection of data exchanged over smartphones..."[62]

For the past two years, the GIMF has maintained a website, the "Technical Center," in Arabic and in English, where users can download encryption software, view tutorial videos, and more. The page states that it is "the technical arm of the Global Islamic Media Front and therefore it is a reference for all Muslims who can benefit from it." It adds, "Without a doubt, it is a small kernel which we will work on to grow larger in the time ahead, Allah the Almighty willing. We ask Muslims upon whom Allah has bestowed the knowledge to know how to promote this website to offer what can benefit their Ummah generally and the Mujahideen particularly. Your brothers in the Global Islamic Media Front Technical Department."

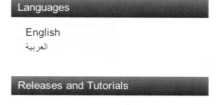

Gimfmedia.com/tech/en/, accessed October 15, 2014.

GIMFmedia.com is hosted in Singapore.

In other examples of the use of encryption, on April 9, 2014, Al-Qaeda's media wing Al-Sahab announced an upcoming open interview with Maulana Asim Umar, head of the Al-Qaeda Shari'a Committee in Pakistan. The announcement said that people could submit questions in Urdu, Arabic, English or Pashto, and included both a Yahoo and a Safe-mail.net email address to where questions could be sent.[63]

Safe-mail, which is an Israeli company, bills itself as "the most secure, easy to use communication system. It includes encrypted mail system with collaboration features and document storage functions. Always accessible at any time from anywhere!" It also states that "Safe-mail is designed to provide maximum security and privacy without any complexity. Banks, law firms, health care, accountants and similar professional organizations will not provide security unless requested by you! Remember, your information is yours only. Your privacy is at risk when you communicate. Do not do business with any of the above unless your valuable information is protected!"

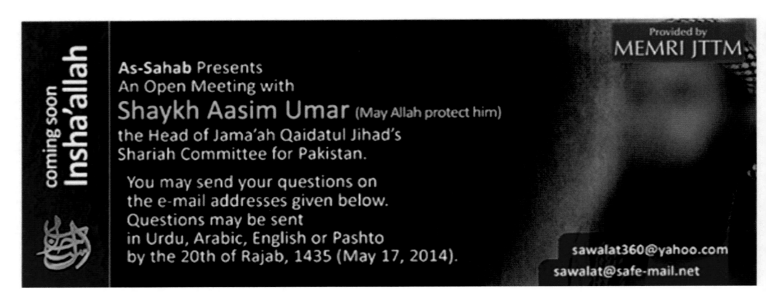

Also, in its first 11 issues, AQAP's English-language magazine *Inspire* provided readers with encryption keys for use by recruits submitting personal information and anyone wishing to contact it. A recent arrest connected with *Inspire* was on March 17, 2014, when 20-year-old Nicholas Teausant of California, who according to the affidavit wanted to bomb the Los Angeles subway system, was taken off an Amtrak train in Washington state at the U.S.-Canada border and charged with attempting to provide assistance to Islamic extremists in Syria. He is being accused of exploring ways to support violent extremist activities since October 2013.[64] The affidavit also notes that he had been active on social media and had posted jihad-related content on his Instagram account. According to the indictment, Teausant frequently referred to Inspire; in one incident he became upset by something in it and texted the informant, "We have to do something now!"[65]

One of Teausant's Instagram posts includes a photo of his computer screen showing him using the GIMF's Asrar Al-Mujahideen ("Mujahideen Secrets") encryption software, which Inspire provided to its readers, along with its own encryption key, for readers to communicate with it (this issue, however, announces that AQAP is suspending its email account). In the post, Teausant explains what it is and writes that he can use it to communicate with the Taliban and Inspire: "Lol its Arsar Al-Mujahideen program it allows you to encrypt messages and safly send them to the taliban repersentitives... also allows me to talk to the creators of 'Inspire.'"

On his Instagram account, Teausant explains what "Mujahideen Secrets" software is and how he can use it to communicate with the Taliban and Inspire.

Islamic State (ISIS) is also using encryption technology in its outreach to English-speaking recruits. The third issue of its magazine *Dabiq* was the first to feature email addresses for contacting it (Dabiq-is@yandex.com, Dabiq-is@india.com, and Dabiq-is@0x300.com), stating, "The Dabiq team would like to hear back from its readers, and for this reason, we are providing email addresses to submit your opinions, suggestions, and questions." It also includes a public encryption key "for those of you who would like to use Asrar Al-Mujahideen" encryption software for secure communication.

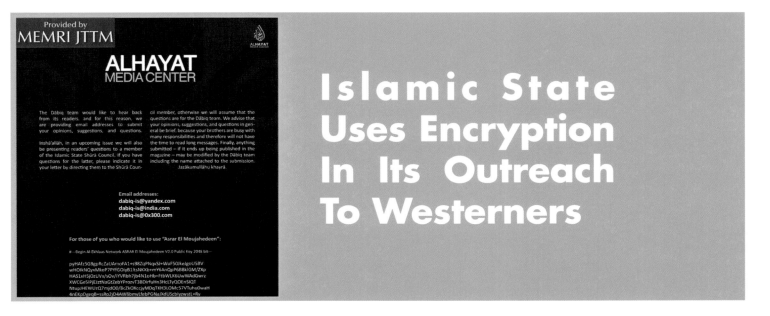

On October 2, 2014, users on pro-ISIS Jihadi Media Platform (alplatformmedia.com) forum announced the launch of a workshop on military electronics, which will be taught by Abu Yahya Al-Shishani ("the Chechen"). In the introduction to his workshop, Al-Shishani stated that he hoped that jihadis could benefit from his experience in the field of electronics, in order to utilize it "in our jihad against the enemies of Allah, His Prophet and the believers." He also reassured jihadis that becoming successful in the field required no substantial prior knowledge in electronics: "Any brother who possesses [knowledge of a] few English terms, and minimal educational level, can be good for this field," he noted.[66]

In addition to encryption technology, jihadi groups are using other means to evade oversight and remain hidden. In late August 2014, several tutorials dealing with jihadis obtaining fake U.S. phone numbers, as well as advice on maintaining security and anonymity online, were posted online. Jihadis sometimes need a working phone number to open or verify social media accounts, messenger apps, and more.[67]

The Taliban's *Azan* online magazine also provides encryption keys for use by recruits submitting personal information and anyone wishing to contact it. Other designated terrorist individuals and organizations – ISIS, AMAF, Lashkar-e Taiba, Jabhat Al-Nusra, and many more – also use encryption technology.

In its new published English-language magazine *Resurgence*, the first issue of which was released on October 19, 2014, Al-Qaeda in the Indian Subcontinent (AQIS) provided an email address and a public encryption key for those looking to contact it. *Resurgence* is a "humble effort to revive the spirit of jihad in the Muslim ummah," according to the magazine. It also encouraged its readers to "participate with us in this effort." In that regard, AQIS welcomed any "advice, feedback, and contributions" from its readers, who were provided with the contact email address resurgencemag@yahoo.com.

Readers were also encouraged to use the Asrar Al-Mujahideen encryption program to contact the magazine, with a public encryption key provided for this purpose. "Take all necessary measures to hide your real identity," the magazine warned, and advised against revealing any personal or sensitive information, even if it was sent securely. "This [i.e. Asrar Al-Mujahideen] software is only a human effort to ensure online security. We do not guarantee that information sent using this software cannot be read by the enemies."[68]

From Al-Qaeda To The Islamic State (ISIS), Jihadi Groups Engage in Cyber Jihad: Beginning With 1980s Promotion Of Use Of 'Electronic Technologies' Up To Today's Embrace Of Social Media To Attract A New Jihadi Generation

III. Statements By Al-Qaeda Leadership On The Importance Of Cyber Jihad

While many of the Al-Qaeda leaders who have pursued its online strategy are now dead, their philosophy remains alive, through videos and articles across the Internet. Among their messages are discussions of the importance of using the Internet for jihad. Anyone skeptical that Al-Qaeda has invested a tremendous amount of effort in the cyber realm need only read what its leadership has said. The following are statements by Al-Qaeda leaders on the importance of cyber jihad.

Osama bin Laden: "The Wide-Scale Spread Of Jihadist Ideology, Especially On The Internet... [Is] A Major Achievement For Jihad"

A May 2010 letter by Osama bin Laden to a "Shaykh Mahmud," that was found inside bin Laden's Abbottabad compound and released by the Combating Terrorism Center at West Point (the Center was given exclusive rights to all content from Abbottabad)[69] reveals just how important Al-Qaeda considers its online activities. In the letter, Osama bin Laden stated: "...[T]he wide-scale spread of jihadist ideology, especially on the Internet, and the tremendous number of young people who frequent the Jihadist websites, [are] a major achievement for jihad..."

Ayman Al-Zawahiri On Cyber Jihad: "A Great Front of Islam"

Al-Qaeda's current leader, Ayman Al-Zawahiri, has on multiple occasions discussed the importance of cyber jihad. In a June 8, 2010 interview with Al-Qaeda's media outlet Al-Sahab,[70] he praised those engaging in online jihadi activity: "[T]o the knights of the jihadi media, I say: May Allah reward you the best reward for your good job in serving Islam. You must know that you are [fighting] on a great front of Islam, and that the tyrants [of our time] are very disturbed by your efforts..."

Adam Gadahn, American Al-Qaeda Spokesman – And Jihadi Media And Cyber Pioneer

Al-Qaeda media leader Adam Gadahn, the American who helped establish and develop Al-Qaeda's media production arm Al-Sahab, has starred in numerous Al-Qaeda video productions and has been a key figure in honing the organization's outreach to the U.S. and the West, with his ability to address an American audience.[71] He was the first member of Al-Qaeda to appear in videos with a computer and a mouse, with an Al-Sahab logo. Al-Sahab was established in 2001; it distributes Al-Qaeda material, promotes global jihad, recruits young Muslims to the cause, and encourages lone-wolf terror attacks in the West. While at first it distributed videos via outlets such as Al-Jazeera, it now disseminates content online through jihadi forums and social media, in English and other Western languages. As the *Washington Post* described it, "Al-Qaeda's core leadership has built an increasingly prolific propaganda operation, enabling it to communicate constantly, securely and in numerous languages with loyalists and potential recruits worldwide."[72]

Al-Qaeda **Media** and **Cyber** Pioneer Adam Gadahn – First Jihadi To Promote **Tech-Savvy** Image In Al-Qaeda Messaging
– Al-Sahab Video, January 2008

Islamic State Of Iraq's Minister Of War: "We Believe Electronic Warfare Is [The] War Of The Future"

Shortly before his death on April 19, 2010, Abu Hamza Al-Muhajir, the Minister of War of Al-Qaeda In Iraq, also known as the Islamic State of Iraq, issued instructions to Al-Qaeda's cyber army, providing them with military instructions: "I urge [you] to [show] interest in the matter of hacking, and to encourage anyone possessing this talent... so [that] we destroy the enemy's [web] sites and infiltrate its military, security and political strongholds...We believe that electronic warfare is [one of] the important and effective wars of the future."

"...the matter of **hacking**... so [that] we destroy the **enemy's [web]sites** and infiltrate its military, security and political strongholds...We believe that **electronic warfare** is [one of] the important and effective wars of the future."
– Abu Hamza Al-Muhajir, 2010

Senior Al-Qaeda Commander to Potential Recruits: We Need "Specialist Cadres," Not "Regular Fighters" In Afghanistan-Pakistan

On July 11, 2010, jihadist websites posted a document written by the late senior Al-Qaeda commander, 'Atiyat Allah Abu 'Abd Al-Rahman Al-Libi, noting that Al-Qaeda has cut back on its recruiting for the Afghanistan-Pakistan arena and is now being highly selective; in an apparent reference to computer skills, he said that the organization is primarily seeking recruits with special expertise.

'Abd Al-Rahman wrote: "...We are in a stage of selection and choosing. We are calling for specialist cadres, which is what the jihad is most in need of, and then regular fighters according to need, and according to what is decided by the commanders of the jihad and those in a position of authority in the jihad. We receive numbers [of mujahideen] bit by bit, choosing and screening. And it is Allah who grants success."[73]

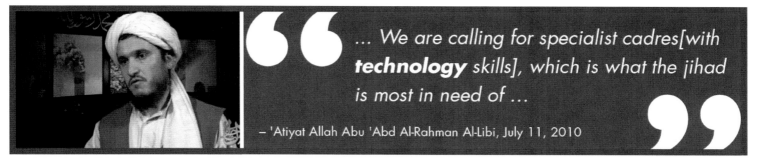

AQIM Publishes First Installment In Electronic Jihad Series

On September 18, 2013, Al-Qaeda in the Islamic Maghreb (AQIM) published an article in what appeared to be a new series focusing on electronic jihad. The document was released via the Twitter account of AQIM's Muslim Africa blog, and was written by one "Abu Musa Al-Shinqiti."

The article, which appears to be a personal initiative by Al-Shinqiti, introduces TOR (The Onion Router), which is a program that provides online anonymity and can bypass online censorship. It presents TOR as a means for providing Muslims with open and anonymous access to AQIM's Muslim Africa blog, and to jihadi websites in general.[74]

Jihadis Discuss Reported Breach Of TOR, Complain That Tech-Savvy Forum Members Have Abandoned Forums For Twitter

An August 12, 2014 MEMRI JTTM report noted that jihadis on the top-tier Al-Qaeda forums Al-Fida' and Shumoukh Al-Islam forums have been expressing their concern over the recent news that the web anonymity project TOR has been breached. On July 30, 2014, TOR announced on its blog that a number of suspicious relays (hopping points through which Internet traffic is encrypted and decrypted) that joined the TOR network at the end of January 2014 were trying to de-anonymize the service's users – especially, it said, those who operated or accessed TOR hidden services. Although the exact implications of the breach remain unclear, the announcement said that "users who operated or accessed hidden services from early February through July 4 should assume they were affected." Jihadis rely heavily on TOR to browse and access jihadi content.[75]

On the Al-Fida' forum, jihadis responded to the news of the breach, and wondered what effect it might have on them. Member Abu Mariam said that as long as jihadis update their TOR browser then they would remain safe. Member Mis'ir Harb2 complained that many forum members had "abandoned us" in favor of Twitter, especially, he said, tech-savvy members whose knowledge on the topic could have been useful. Abu Jihad Al-Muhandis, an admin of the modern technology section on Al-Fida', also noted that "unfortunately, there are technical brothers [who] abandoned Al-Fida and Shumoukh [Al-Islam in favor of] Twitter," and prayed for their return.

On the Shumoukh Al-Islam forum, jihadis expressed concerns similar to those expressed by Al-Fida' members. Muraqib9, a supervisor of the technical section on the forum, recommended that jihadis update their TOR browser to the latest version. He also recommended that jihadis use trusted encryption software like Asrar Al-Mujahideen and 'Amn Al-Mujahid, in addition to using TOR.

A few months later, in October 2014, jihadis were back using TOR regularly. For example, on October 13, 2014, the Twitter account for Warshat Fursan Al-Nashr, a subsidiary of the Al-Battar media company – the mouthpiece of the Shumoukh Al-Islam forum – posted a link to a tutorial on installing it on Android devices.[76]

Al-Qaeda in the Indian Subcontinent (AQIS) published web links for jihadi fighters to download a specially modified TOR browser on November 14, 2014. Details of the modified TOR software were released via Twitter. On the same day, a tweet from the account linked to details of the modified TOR for Al-Qaeda militants to be downloaded from justpaste.it, and a November 4 tweet advised jihadis to download the Orbot software for mobile phones from Google Play. Yet another tweet urged jihadis to use a Russian email service provider. AQIS also retweeted a tweet from its spokesman Usama Mahmood, which gave details on the Asrar Al-Mujahideen encryption key for potential jihadists to contact.

AQAP Commander Calls On Followers To Learn Tools for Cyber Jihad From The Internet

A video of Al-Qaeda in the Arabian Peninsula (AQAP) commander Qassem Al-Rimi, released April 10, 2012, shows him urging his followers to embrace cyber jihad.[77] In the video, Al-Rimi says, "I call on Muslims in general to get the correct and clear information via the jihadi websites on the Internet, [since] our brothers [on it] speak only the truth." The video then goes on to show viewers how to download a TOR Browser Bundle in order to protect their identities online. The use of this software by jihadis has been the subject of much speculation following the recent NSA controversy, as it gives jihadists a way around the now-publicized methods used by the U.S. government to monitor terrorist threats.

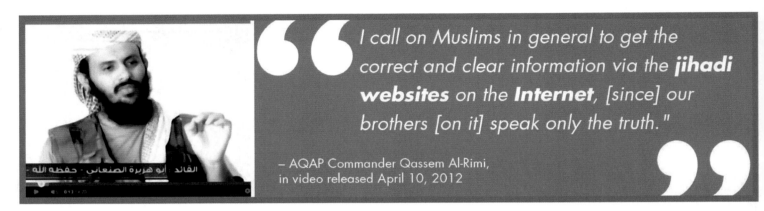

AQAP Deputy: Online Jihad Is A Great Front And Part Of The Coming Phase

In an essay penned under the pseudonym "Abu Asma' Al-Kubi" titled "Be a Mujahid," and released on February 22, 2012, just over a year before he was killed in a U.S. drone strike in Yemen in the summer of 2013, AQAP deputy leader Sa'id Al-Shihri discusses cyber jihad efforts, and states: "Prepare for the coming phase, and let all of us be media men if we are not attached to a work, for there might come a time when the jihadi media work might be difficult."

He continued: "The Ummah still requires incitement and direction, for our enemies never tire from spreading their thoughts or lures in our societies day and night while they are upon falsehood, yet they are at their peak in spreading their falsehood. So it is mandatory upon us to keep our spirits higher than theirs, and that we get involved in the media because that is the precursor to conquest and military work. So work hard on this great front, and every person based on his ability and knowledge, those who publicize, let them publicize, and there must exist a harmony in their work so that our psychological conquest reaches all people, the supporters as well as the haters and so forth."[78]

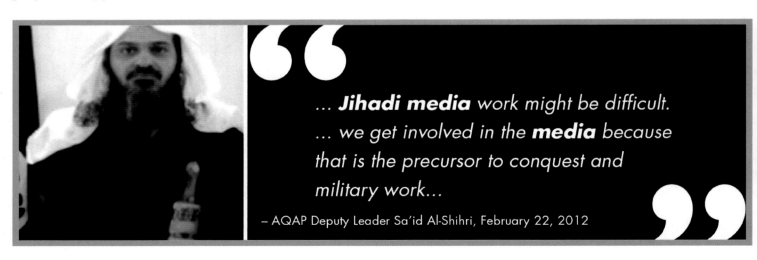

Al-Qaeda Deputy Leader and Former Osama bin Laden Secretary Nasir Al-Wuheishi Calls For Volunteers to Join Al-Qaeda Via Encryption Software

The October 2009 issue of Al-Qaeda in the Arabian Peninsula's (AQAP's) *Sada Al-Malahim* magazine included information for jihadists seeking to get involved in cyber jihad, in particular to expand their intelligence database through crowdsourcing – outsourcing tasks to an undefined group of people through an open call – as well as an appeal to readers to send in information they had collected on political, military, and security officials and on Western military targets.

The magazine included an essay by Nasir Al-Wuheishi, the Yemeni-born Emir of AQAP, former personal secretary to Osama bin Laden, and recently named deputy of Ayman Al-Zawahiri, which discussed Al-Qaeda's use of encryption software and outreach efforts. He wrote: "For our part, we will make contact with anyone who wants to wage jihad with us, and we will guide him to a suitable means to kill the collaborators and the archons of unbelief – even in his bedroom or workplace. Anyone who wants to give support to [AQAP's] operational side and to give tithes [to the organization] can contact us through a special email [set up] for this purpose, using the 'Mujahideen Secrets' software and employing the proper security measures..."[79]

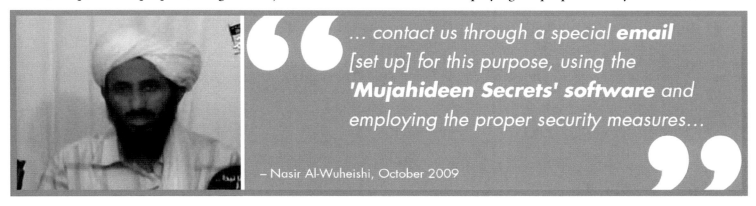

Anwar Al-Awlaki – The "Bin Laden Of The Internet": The Internet Is A Great Medium For Spreading Jihad

One Al-Qaeda leader who understood more than anyone the importance of Al-Qaeda's embrace of the Internet was the American-born Sheikh Anwar Al-Awlaki, who earned the nickname "the bin Laden of the Internet."[80] Al-Awlaki put the Internet at the center of what he thought Al-Qaeda's future strategy should be. Before many Westerners knew of him, he had his own website, where he published sermons and articles.[81] He also provided contact information and email accounts for followers to contact him – including "underwear bomber" Umar Farouk Abdulmutallab and Fort Hood shooter Nidal Malik Hasan.

In a May 23, 2010 interview with Al-Sahab, Al-Awlaki's frustration at the impact of the shutdown of his website by U.S. authorities was clear:[82] "They shut down my website following Nidal Hasan's [Fort Hood] operation. I had posted an article of mine in support of what Nidal Hasan did, and so, they shut down my website. Then I read in the *Washington Post* that they were monitoring my communications. So I was forced to stop these communications."[83]

On January 5, 2009, Al-Awlaki published an essay titled "44 Ways of Supporting Jihad," based on an article in Arabic by Muhammad Al-Salem, "39 Ways of Supporting Jihad." Al-Awlaki's essay, which has been spread widely over jihadi forums and YouTube, includes a section titled "WWW Jihad," in which he states: "The Internet has become a great medium for spreading the call of Jihad and following the news of the Mujahedeen. Some ways in which the brothers and sisters could be 'internet mujahedeen' is by contributing in one or more of the following ways: establishing discussion forums that offer a free, uncensored medium for posting information relating to Jihad; establishing email lists to share information with interested brothers and sisters; posting or emailing Jihad literature and news; setting up websites to cover specific areas of Jihad, such as: mujahedeen news, Muslim POWs, and Jihad literature."[84]

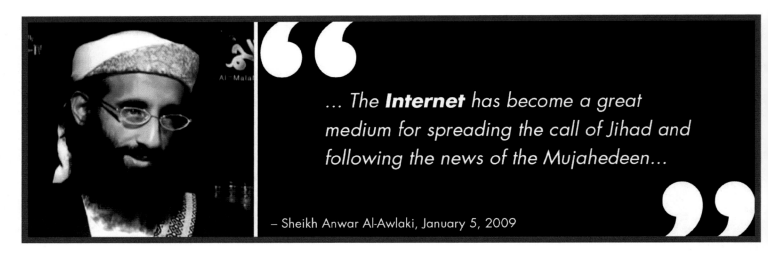

Inspire Editor Samir Khan Warns The West That Al-Qaeda Cyber Activists Are Studying Internet Security, Praises The Impact Of Made-In-Afghanistan Jihadi Videos That Are Distributed In The Streets Of London And California

Al-Awlaki's protégé, American AQAP member Samir Khan, who was editor of Al-Qaeda's English-language online magazine *Inspire*, was also a leading figure in Al-Qaeda's push for cyber jihad; he was killed alongside Al-Awlaki in a September 30, 2011 U.S. drone strike in Yemen.

In an article titled "The Media Conflict" in Issue VII of *Inspire,* which was released just three days prior to his death, Khan discussed the importance of 'media jihad' as a component in the war against the U.S., equating it with an actual attack on it: "A powerful [jihadi] media production is as hard hitting as an operation in America." He continued: "There were namely three things that the brothers focused on in their media efforts: quality and content of productions, Internet security and a media dissemination strategy. While America was focused on battling our mujahidin in the mountains of Afghanistan and the streets of Iraq, the jihadi media and its supporters were in fifth gear. Thousands of productions were produced and dispersed to both the net and real world. Something that was produced thousands of feet above in the mountains of Afghanistan was found distributed in the streets of London and California. Ideas that disseminated from the lips of the mujahidin's leaders were carried out in Madrid and Times Square."[85]

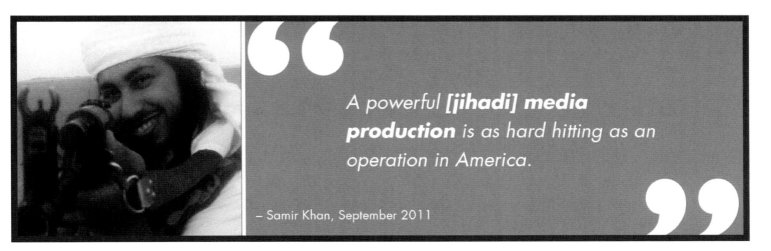

A powerful [jihadi] media production is as hard hitting as an operation in America.

– Samir Khan, September 2011

U.S.-Designated Global Terrorist Abu Adam Al-Almani: Al-Qaeda's "Professional Media Work" In German And English "Reached Us In Germany"

The importance of Al-Qaeda media efforts was emphasized by U.S.-designated global terrorist Abu Adam Al-Almani in an August 2013 interview with the Taliban's online English-language magazine *Azan*. In the interview, he recounted how online Al-Qaeda videos had helped in his radicalization, inspiring him to move to Yemen and meet with the late Al-Qaeda leader Anwar Al-Awlaki. He said that Al-Qaeda's "professional media work" in German and English "reached us in Germany," including lectures by Al-Qaeda leaders Ayman Al-Zawahiri and Abu Yahyah Al-Libi, and that "therefore, we started to follow the right Manhaj (path)." He pointed out that a particular Osama bin Laden interview is used by "the brothers until today for their videos," because bin Laden "explains every step regarding what they did."[86]

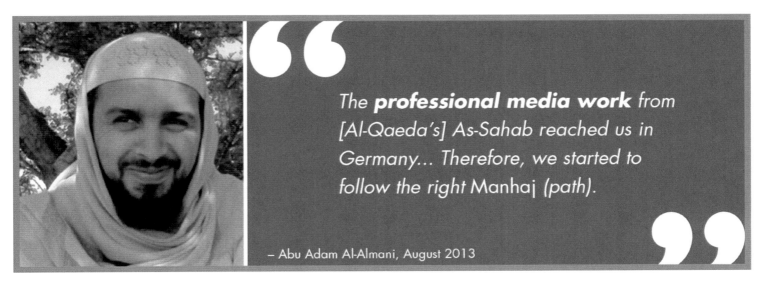

The professional media work from [Al-Qaeda's] As-Sahab reached us in Germany... Therefore, we started to follow the right Manhaj (path).

– Abu Adam Al-Almani, August 2013

IV. A Major Shift In Online Jihad: From Forums To Social Media

The period following the killing of Al-Awlaki and his protégé Samir Khan marked a shift for Al-Qaeda and its affiliates: From relying solely on a handful of main forums for distributing its official releases and other content (down from thousands of sites which previously existed), it was now increasingly relying on U.S. social media companies. A decade of monitoring jihadi websites shows that at this point there began a significant decline in traditional, closed jihadi forums that require registration to an explosion of jihadis and jihadi groups on social media, where content is out in the open and easily accessible.

Following Killing Of Al-Awlaki and Khan, Al-Qaeda Cyber Activists Promise To Spread Online Jihad And Raid Facebook And Twitter

Immediately following the killing of Anwar Al-Awlaki and Samir Khan, online jihadis promised that they would avenge them by spreading online jihad, by flooding U.S. websites and social networks.

A Shumoukh Al-Islam forum member who teaches a course in Rapidleech, a script that allows users to share files using various online hosts, wrote in English: "Together for Islamizing [the] U.S.A. I want to see the American forums, websites, YouTube channels, and Twitter full of Sheikh Anwar Al-Awlaki's [sic] lectures and videos. It will be a curse chasing the Americans and their dogs." He added in Arabic: "To the lions of uploading and [online] distribution, to the students of Anwar Al-Awlaki... You have graduated from the Rapidleech course. Do you remember the pledge we took in it? 'I entrust you with supporting your brothers, the mujahedeen, with this knowledge. At the very least, we shall spread their actions and the truth... You are the hope of the ummah, you are the hope of jihad... Let it be a vengeance raid for our sheikh, Al-Awlaki. We want a raid of every American forum, every Facebook page, every Twitter account...'"[87]

American Al-Qaeda Spokesman Adam Gadahn: "We Must Make Every Effort To Reach Out To Muslims Through New Media Like Facebook and Twitter"

In a March 1, 2013 interview with *Inspire,* Adam Gadahn, who heads Al-Qaeda's media wing Al-Sahab and who for years has appeared in videos at his desk with his computer, spoke directly to Western supporters of Al-Qaeda and to potential recruits. Noting the importance of American social media companies, he said: "This is your day, so rise to the challenge and become a part of history in the making... we must make every effort to reach out to Muslims both through new media like Facebook and Twitter... and we should fully acquaint ourselves with both the people to whom we are reaching out, as well as the methodology and cause to which we are inviting them, so that we are able to hone our methods, refine our techniques, and spread our message in an intelligent and educated fashion accessible to all sectors, sections, levels and factions of the ummah."[88]

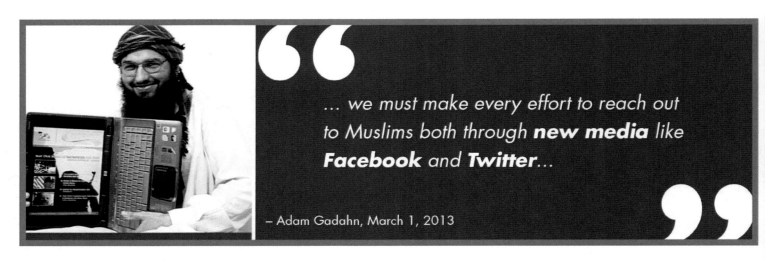

From Al-Qaeda To The Islamic State (ISIS), Jihadi Groups Engage in Cyber Jihad: Beginning With 1980s Promotion Of Use Of 'Electronic Technologies' Up To Today's Embrace Of Social Media To Attract A New Jihadi Generation

Bali Bombing Mastermind: "This Is The Internet Era, There Is Facebook, Twitter, And Others"

Umar Patek, who masterminded the October 2002 bombing in Bali that targeted Westerners in several bars and night clubs in Bali, Indonesia which left 202 dead, warned on June 7, 2012 during his trial: "...For those who do not know how to commit jihad, they should understand that there are several ways of committing jihad." He added: "This is not the Stone Age... This is the Internet era, there is Facebook, Twitter and others."[89]

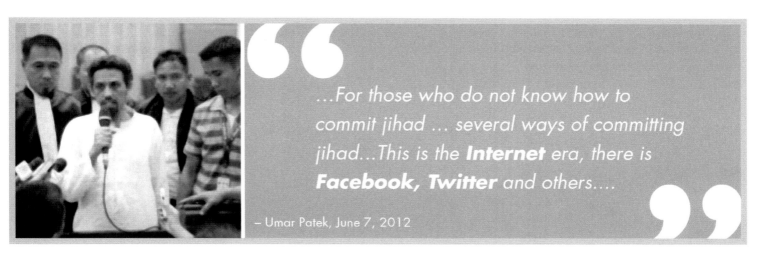

Taliban Spokesman Praises Impact Of Online Jihad And Use Of Facebook And Twitter

Just six months prior to the killing of Al-Awlaki and Khan, Abdul Sattar Maiwandi, a Taliban commander and leader of the Taliban's official website, said in an interview published February 18, 2011 by Al-*Somood*, the Taliban's Arabic-language magazine: "... We are also active on Facebook and Twitter, where we publish the news every day and reach thousands of people."[90]

Qari Yousaf Ahmadi, spokesman for the Islamic Emirate of Afghanistan, the shadow government of the Taliban, said in an April 15, 2012 interview with the Saudi London-based daily *Al-Sharq Al-Awsat*: "Praise be to God; I use computers and have accounts on Facebook, Twitter, and YouTube." Ahmadi went on to explain: "The media is a basic and important part of the ongoing war between us and the occupation enemy. [Winning] the media war means, perhaps, winning more than half of the war...as long as we are in a state of war, we will use all modern means [i.e. media technologies] available, and acquire all possible expertise."[91]

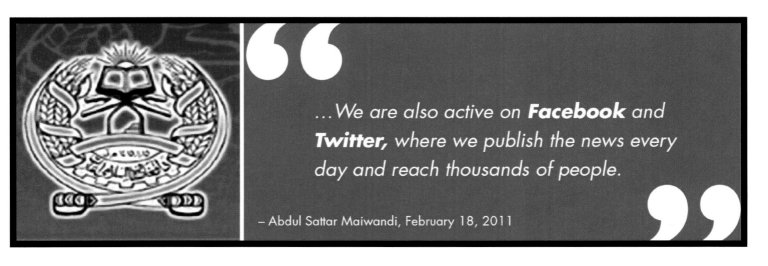

Jihadi News Agency "Kavkaz Center," Affiliated With Designated Terrorist Organization "Caucasus Emirate," Calls For Followers To Use Twitter and Facebook

On January 27, 2012, Kavkaz Center posted on its website an announcement that urged its followers to "[c]reate accounts on Twitter, Facebook and share the information... Modern technologies make it possible for anyone who genuinely cares about the Mujahideen not to stand aside [but to] render them real help, becoming a participant in the fight of truth against the lie."[92]

Kavkaz Center is closely affiliated with the Caucasus Emirate (CE), a jihad organization operating in Russia's Caucasian republics, and has posted videos featuring messages from and interviews with its leaders. The website has also featured exclusive information on developments in the CE, which could only have been obtained through close ties with CE operatives.[93] In May 2011, the CE was designated a Terrorist Organization by the U.S. State Department under Presidential Executive Order 13224;[94] a year previously, its leader Dokku Umarov was designated a wanted foreign terrorist.[95]

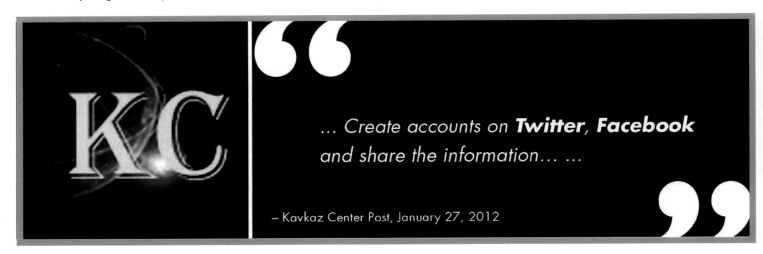

Former Guantanamo Detainee And Al-Qaeda Cyber Activist Praises Online Jihad: "Especially Through Twitter And Facebook"

On October 8, 2012, Walid Muhammad Hajj, a former Guantanamo detainee from Sudan who fought in Osama bin Laden's 55th Arab Brigade, tweeted: "The road to paradise is surrounded by hardship and however there is a way out [that is] by jihad through the net especially through Twitter and Facebook..." Hajj, who joined Twitter (WaleedGaj2002) on June 13, 2012, has as of this writing posted over 7,800 tweets, has nearly 69,000 followers, and is following over 50 well-known jihadis and jihadi groups. Hajj often tweets pro-Al-Qaeda and anti-West statements.

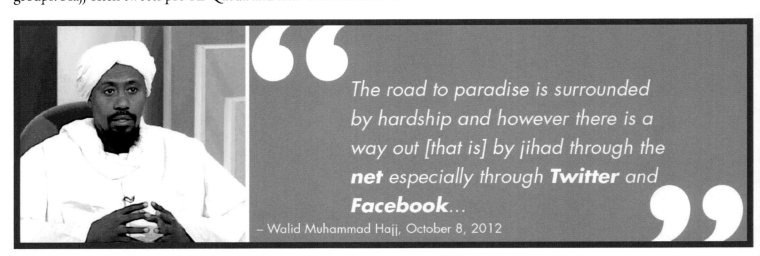

V. Western Social Media Companies – At The Heart Of Al-Qaeda, ISIS, Other Groups' Jihadi And Outreach Efforts

U.S.-Based Social Media Companies – The Engine Of Jihad Today

Social media are increasingly instrumental in spreading Al-Qaeda's ideology to the younger generation, now in their 20s or even younger, who have grown up watching video clips on YouTube and for whom social media are an integral part of life. Al-Qaeda and its supporters have now infested YouTube, Twitter, Facebook, Instagram, and Flickr, and are now spreading to newer media as they develop – Ask.fm, Kik, Friendica, and, most recently, VK.com, Diaspora, JustPaste.it, and SoundCloud. These same users are also utilizing apps that are available on Google Play and iTunes for Apple.

While jihadi forums have been limited to registered users and are often password-protected, social media has created a more open flow of jihadi ideology in real time; in many instances now, information in the jihadi world appears on social media before it is released by the forums, and many jihadis have openly discussed social media as a game changer. Furthermore, when jihadis expand into a social media platform where they have not previously had a presence, they devote a great deal of effort to telling their fellow jihadis about the new platforms and to explaining in detail how to use them effectively, with video tutorials and more.

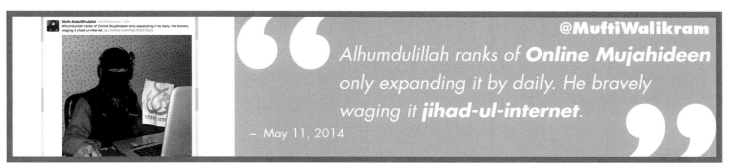

Illustrating Al-Qaeda's awareness of just how vital social media are today, in a December 26, 2013 video by Al-Qaeda's media wing Al-Sahab, Warren Weinstein, the American aid worker who held by Al-Qaeda since his kidnapping in Pakistan in August 2011, urged the American public, as well as journalists and writers, to use them to pressure the White House to secure his release.

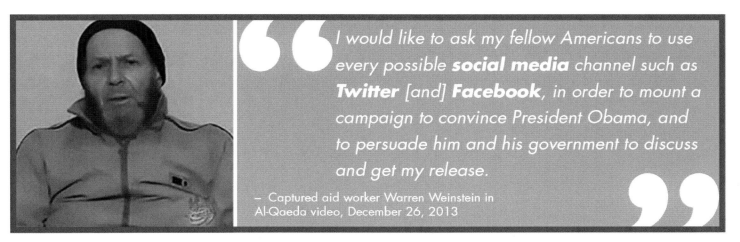

In the 13-minute video, which was sent to a U.S. media organization, Weinstein states, possibly reading from written text: "I would like to ask my fellow Americans to use every possible social media channel such as Twitter [and] Facebook, in order to mount a campaign to convince President Obama, and to persuade him and his government to discuss and get my release."[96]

The San Francisco-Based Internet Archive – Platform For Uploading And Downloading Al-Qaeda Content: Fast, Free, And Unobstructed For Terror Organizations

In addition to social media, the San Francisco-based Internet Archive (archive.org) has in recent years become an important platform for Al-Qaeda content, including videos; jihadis are uploading to and downloading from the Internet Archive on a daily basis. Many Al-Qaeda-affiliated websites now include links to Archive.org with posts of new material online – including the most recent speeches by the Al-Qaeda leaders and many other productions by Al-Qaeda's Al-Sahab media company. This content is often reposted on other websites. Members of the leading jihadi forums also frequently instruct their readers to use the Internet Archive; for example, on July 20, 2011, a member of the major jihadi forum Shumoukh Al-Islam gave readers detailed steps for uploading material there.[97] Using it is quick and easy, and the Internet Archive is doing nothing to stop its use by jihadis. At one point, when the main forums were beginning to be taken down and were no longer reliable, Al-Qaeda and its offshoots used the Internet Archive as their own official forum. Today, it is common for online jihadis to post tweets with links to documents and videos on Archive.org.

Twitter.com/truthsMaster; Twitter.com/s7bhijratain

YouTube – The Internet's Primary and Rapidly Expanding Jihadi Base

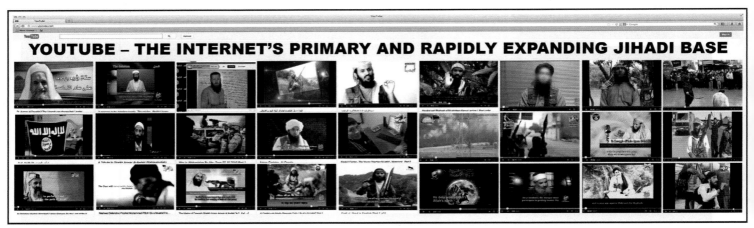

For over four years, YouTube has been very extensively used by Al-Qaeda and its offshoots – so much so that it has not only surpassed but has replaced the use of websites administered by the jihadis themselves.[98] A number of Al-Qaeda sympathizers involved in terrorism cases maintain active YouTube pages. Additionally, every major video released by Al-Qaeda to its affiliates is uploaded to YouTube as soon as they are posted on forums.

On many of the Al-Qaeda forums there have been ongoing discussions of the importance of YouTube, and training has been provided for using it most efficiently. In one example of this, in August 2011, the Shumoukh Al-Islam forum announced that it was offering a course in uploading content to YouTube and the Internet Archives for uploading to YouTube, "to safeguard the mujahideen's legacy." The course announcement stated: "My brothers, surely you are aware of the importance of uploading the mujahideen's productions to the [Internet] Archive [Archive.org], Youtube, and other uploading sites, seeing that the links to the mujahideen's original productions have been deactivated..."[99]

Twitter – Hashtag Jihad And Fundraising For Jihad

Just as important to Al-Qaeda, its allies, and its online followers is Twitter, which is now being used by terrorist organizations and their media outlets, and the number of their online followers have grown exponentially in recent years.

These organizations include many officially recognized by the West and U.S. government as designated terrorist entities, including Al-Qaeda affiliates. The main jihadi forums – Shumoukh Al-Islam, Al-Fida', and AMEF, as well as senior writers on these forums – all have at one time had Twitter accounts. Other terrorist entities tweeting include the Taliban (alemarahweb and ABalkhi); Hamas (hamasinfo) and its military wing Al-Qassam Brigades (AlqassamBrigade); Hizbullah and its Al-Manar TV (almanarnews); and countless others.

When the MEMRI JTTM first began monitoring jihadis on Twitter, there was only a small handful of them. This number quickly grew to hundreds, and then thousands; there are now tens of thousands of such accounts. MEMRI has extensively researched how these organization use Twitter to promote their agendas, spread their messages, call for attacks against American and Western interests, recruit new members and build their audience of sympathizers, raise funds, and other purposes.[100]

One example highlighting how online jihadis and terrorists such as Al-Qaeda and Al-Qaeda figures have increasingly been using Twitter is the Ansar Al-Mujahideen Forum's (AMAF's) posting of "important instructions" for jihadis for using the service. On May 7, 2012, the forum announced that since it had grasped the crucial role jihadi media plays in the battle between Islam and its enemies, it was using all legitimate means to support Islam, including a new Twitter account (as_ansar) that it had opened the previous month. The "important instructions" to jihadis were to follow AMAF's Twitter account and to retweet its posts; they also included an explanation of how to use hashtags.[101]

The following month, a message posted on the Ansar Al-Mujahideen Arabic Forum (AMEF) on June 5, 2012 announced that "Asad Al-Jihad2," a prominent writer on jihadi forums, had opened a Twitter account (@AsadAljehad2). The announcement stated that Twitter serves as a "very important" platform for delivering personal messages, both privately and publicly, and that it therefore enables users to overcome the "media barriers" set in place by the enemies of Islam aimed at stopping "[those who] possess the truth" from communicating with the masses of the *ummah*.[102]

Another example is the July 21, 2013 official launch by users of the top jihadi Al-Fida' forum of "the jihadi media brigade: Al Battaar Media." Al-Battaar's official page, @Al_Bttaar, began tweeting July 17, 2013, and as of this writing had over 27,500 followers and had posted over 3,200 tweets. Its purpose is to spread jihadi content online; it provides a Gmail address as contact information.

Hacking, spamming, and takeover attempts of Twitter accounts have been jihadi Twitterverse staples in recent years. For example, one Twitter campaign, with the handle @Spam_campaign, was aimed at coordinating spam attacks against anti-jihadi Twitter accounts designed to shut them down.

Twitter.com/AL_Bttar, Twitter.com/Spam_campaign, Accessed December 23, 2013

The campaign's administrators post links to other accounts that "defame the mujahideen" and ask jihad supporters to help shut them down by filing a complaint against a targeted account using Twitter's "report [name of account] for spam" option. On November 30, 2013, for example, the Jaish Al-Ansar campaign urged jihad supporters to file a spam complaint against the U.S. State Department Digital Outreach Team's Arabic account (@ DSDOTAR): "Knock it down, oh lions of monotheism!" The call received 27 retweets, and nine users reported that they had actually filed the complaint.

Images created to express support for jihad and jihad groups, such as the commonly seen one below showing a bullet, a pen, and a thumb drive representing online jihad, are circulated frequently on Twitter; this one has the ISIS stamp and states "There is a different form of jihad... what's important is [that you] not abandon your place [i.e. position]."

A Bullet - A Pen - A Thumbdrive: "There is a different form of jihad...what's important is [that you] not abandon your place [i.e. position]."

Twitter is also widely used for fundraising for jihad. For example, on February 26, 2014, Sheikh 'Abdallah Al-Muhaisni, a Saudi cleric who has joined the mujahideen in Syria, launched a Twitter fundraising campaign (@Jahd_bmalk) to buy ammunition for the "Islamic brigades" fighting in Syria. According to various tweets from the account, over 26,000 riyals have been donated thus far. A previous campaign was titled "Participate in Jihad with your Money."[103]

A campaign underway on Twitter in the spring of 2014 stated, "Support the Mujahideen with financial contribution via the following reliable accounts" and provides contact information on other Twitter accounts.

Twitter.com/Jahd_bmalk and closeup of fundraising image; Twitter.com/Khalid_Maghrebi

In previous Twitter fundraising campaigns, photos of donations such as stacks of gold bars, luxury cars, and so on were circulated, along with photos of the weapons purchased with proceeds from their sale.

Friending Al-Qaeda On Facebook

A growing number of Al-Qaeda affiliates and other designated terrorist organizations, as well as online activists who support terrorist organizations, have been active on Facebook; while some of their accounts have been shut down, they often return. Some leading pages are those of the Al-Qaeda offshoot ISIS's media company Al-Furqan,[104] Al-Qaeda in the Islamic Maghreb (AQIM),[105] and Jihad Umma, which reports from jihadi websites.[106] Other organizations on Facebook include the Taliban, Hamas, Hizbullah, and individual jihadi leaders and sheikhs.

Mustafa Maya Amaya, the suspected leader of an international jihadi recruitment ring that was broken up by Spanish and Moroccan security forces in March 2014, used Facebook as a recruitment tool. Jihadis recruited by him, or associated with him – all of whom are as of this writing fighting in Syria – are linked to him on Facebook.[107]

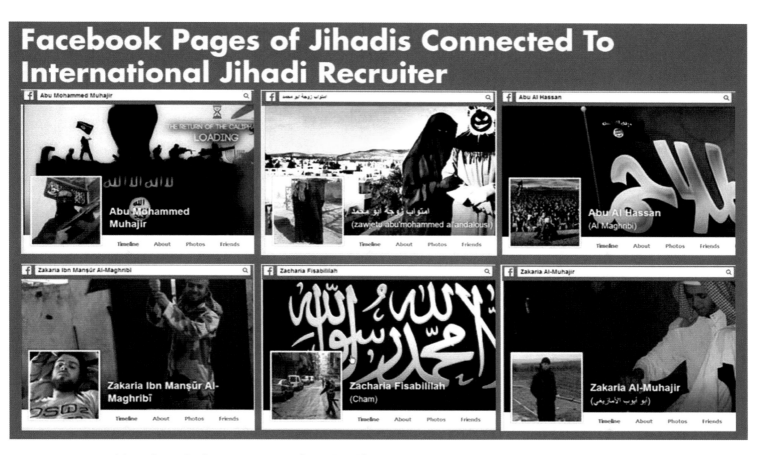

In 2012, one notable online jihadist wrote: "This [Facebook] is a great idea, and better than the forums. Instead of waiting for people to [come to you so you can] inform them, you go to them and teach them!" Other stated goals are: reach the wide base of Muslims who [use] Facebook, encourage brothers to devise new online media in support of jihadi media, form a solid base on Facebook and shed light on it as a medium for reaching people. Move from an elite society ([on] jihadi forums and websites) to mainstream Muslims, [encourage] their participation, and interact with them, advance media operations and encourage creativity, innovation, flexibility, and change. Reach large [numbers] of Crusaders, broadcast the losses of their armies, expose the lies of their leaders, and call Muslims to jihad."[108]

One of the first reported instances of the jihadist message on infiltrating Facebook was in September 2008, after jihadi forums had been temporarily shut down and the forum members began using Facebook as another venue for convening and spreading jihadist content.[109] The first phase of social network jihad consisted mostly of "Facebook raids," or campaigns aimed at disseminating jihadi propaganda through existing Facebook channels. The second phase has seen jihadists establishing a permanent Facebook presence by creating groups and virtual communities affiliated with leading jihadist websites. In this way, numerous features offered by Facebook have been used to promote the ideology of jihad and to spread its messages.[110]

On December 20, 2013, jihadi groups announced via several Facebook accounts and online forums the launch of a question and answer session with German jihadi fighters based in Waziristan, which is the Taliban and Al-Qaeda stronghold in the Pakistani tribal region. A message was posted in Urdu and English on the Jamia Hafsa Urdu Forum Facebook page: "We are getting ready for Q&A session/interview with German brothers [i.e. fighters] in Waziristan. If you have any question please send your questions in inbox." As of December 23, this jihadi account on Facebook had 2,312 followers.[111]

Also, Nasser Al-Ansi, a Yemeni Al-Qaeda in the Arabian Peninsula (AQAP) figure, has been recruiting individuals from the Gulf region, especially from Saudi Arabia and Kuwait, via Facebook. Al-Ansi also communicated with those recruits via email and over the phone. A December 2013 report said, citing a Yemeni security source, that Al-Ansi had good "persuading" skills. Al-Ansi may be deputy to AQAP leader Nasser Al-Wuhaishi, whose deputy Sa'id Al-Shihri was killed in a U.S. drone strike earlier this year..[112]

Al-Qaeda and its online sympathizers are also actively experimenting with apps and other features on Facebook. In March 2013, a member of the leading jihadi forum Al-Fida', Sayf ibn Dhi Yazan, announced the release of what he said was the first jihadi application for Facebook. He said that the application would facilitate the direct release of jihadi productions to subscribers' pages.[113] In his announcement, Yazan called on forum members to support his project so as to enable him to "complete this stage and move on to another one." The Facebook page to which Yazan linked is filled with jihadi propaganda videos and interviews, most of them recent releases by AQAP, Islamic State of Iraq (ISI), and more. The page had 57 "likes" on its first day.

In addition, jihadi groups are using Facebook for recruiting purposes. In a recent example, a Gaza Al-Qaeda operative recruited three Palestinians over Facebook and Skype, according to January 22, 2014 media reports. The three were planning the organization and execution of attacks in Israel, including a simultaneous double suicide bombing at the U.S. Embassy in Tel Aviv and the International Convention Center in Jerusalem. One suspect had received computer files containing virtual training courses in bomb manufacturing from the Gaza Al-Qaeda operative, and a second said that he too had learned to manufacture bombs online. Israeli authorities announced that the three had been arrested.[114]

Designated Terrorists And Terrorist Organizations Online: Maintaining Official Websites, Using Google Blogspot, Using Yahoo Server, Launching Internet Radio Stations

Designated terrorists using various media platforms abound. A few prominent examples are: Omar Abd Al-Rahman (the Blind Sheikh), who is serving 15 life sentences in the U.S., and who has an official website operated on a Chicago-based ISP as well as accounts run by his son on YouTube, Twitter and Facebook[115]; Specially Designated National Hafiz Muhammad Saeed, founder of the U.S.-designated Foreign Terrorist Organization Lashkar-e-Taiba (LeT) and emir of its charity arm Jamaatud Dawa, who is also wanted by India for his role in the planning of the 2008 Mumbai terror attacks, and who is also active on YouTube and Twitter;[116] Syrian Jabhat Al-Nusra and its leaders, and countless others, on YouTube, Twitter, Facebook.

A trailblazer in jihadis' embrace of social media was FBI Most Wanted Terrorist Omar Hammami, an American commander in the Al-Qaeda affiliate in Somalia, Al-Shabaab Al-Mujahideen, who was killed by his organization in Somalia in September 2013. One of the first Western jihadis to gain a worldwide following, Abu Mansour Al-Amriki, as he was known, he was prolifically active on Facebook, Twitter and YouTube; he even released his autobiographical ebook *The Story of an American Jihaadi – Part One* via his YouTube account. In the book's introduction, he noted that he had decided to pen his memoirs due to the "unpredictable nature" of his life in the land of jihad, and in accordance with the advice of Al-Qaeda military strategist and ideologue Abu Mus'ab Al-Suri, who had underlined the importance of documenting one's history. Such documentation was particularly important today, he said, now that the Internet has made it simple to record events and spread reports of them. Today, a year after his death, his Twitter and Facebook accounts remain online.[117]

Al-Qaeda in the Islamic Maghreb (AQIM) announced, on May 2, 2013, the launch of "the Muslim Africa Blog" aimed at persuading Muslims in Africa to "return to their Islamic identity, which the unbelieving colonizers have worked hard to keep them away from." The blog, on Google's social media blogging platform Blogspot.com, has corresponding Twitter and Facebook pages.[118] Additionally, the Al-Twahid Wal-Jihad Movement in West Africa, a Salafi-jihadi organization and Al-Qaeda affiliate, used a Yahoo server to launch a jihadi forum offering training courses in bomb making and booby traps.[119]

Muslim Africa blog, on Google's Blogspot. Accessed December 18, 2013; February 18, 2014 tweet stating that the broadcasts are experimental.

On February 18, 2014, the twitter account of Syria Al-'Aan (@Syria_now_1) reported that the Islamic State of Iraq and Al-Sham (ISIS) had launched radio broadcasts of religious lessons and statements by ISIS leader Abu Bakr Al-Baghdadi (now Islamic State Caliph), in the Syrian town of Al-Raqqa. According to one tweet, the radio broadcasts, which can be heard on 99.9 FM, are at this stage experimental.[120] Previously, in January 2013, Jihad Al-Umma launched what it claimed was the first jihadi Internet radio station, Radio Fajr Al-Jihad.[121]

VI. Social Media In The Syria And Iraq Conflict

The use of social media in the conflict in Syria and Iraq highlights the global jihad movement's complete dependence on the Internet and on U.S.-based social media companies – and is the template for the future of jihad. A large number of foreign jihadis who have gone to Syria and Iraq use Facebook, YouTube, Twitter, Flickr, Tumblr, Ask.fm, SoundCloud, and other social media to share experiences and updates, plan attacks, fundraise, keep in touch with family, friends, and followers, and answer questions.

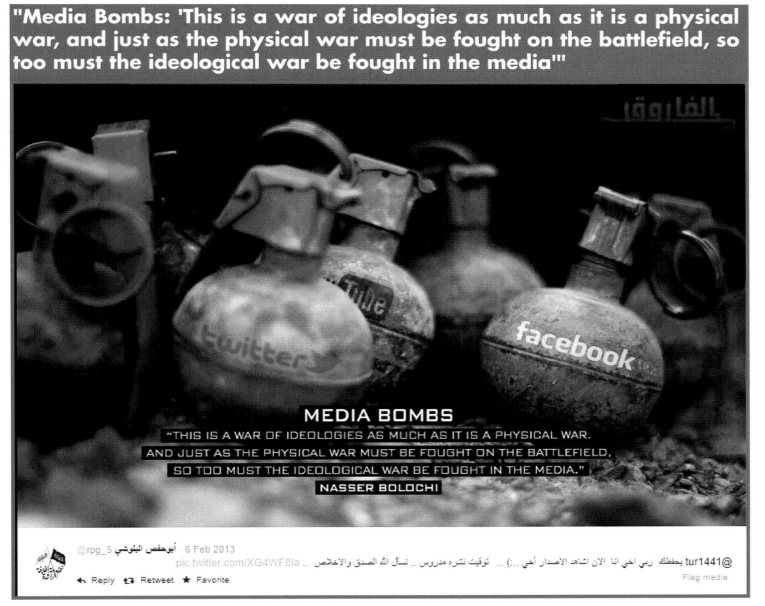

"Media Bombs: 'This is a war of ideologies as much as it is a physical war, and just as the physical war must be fought on the battlefield, so too must the ideological war be fought in the media' – Nasser Bolochi" (Twitter.com/rpg_5/status/299398930255470592/photo/1)

Many of these foreign jihadis are in fact active members and writers – even prominent writers – on the Al-Qaeda and other jihadi forums; news that one or another of them has been "martyred" is frequently tweeted and otherwise disseminated via social media. Unless noted otherwise, the jihadis discussed in this report were alive and operating at the time of writing; it also should be noted that their social media accounts generally remain online after their deaths.

"Behind the Scenes" of jihad – Image of "Admin" on computer. Twitter.com/JMuhajirah/status/441626800909602816/photo/1; @Abualbawi, accessed December 23, 2013; Twitter.com/HalaJaber

German rapper-turned jihadi Abu Talha Al-Almani, formerly known as Deso Dogg, appears to be involved with the ISIS media arm Al-Hayat, following his April 2014 video interview in which he announced that he had decided to quit active fighting and assume a new role in the ISIS dissemination mechanism. He said in the interview: "That's why I pledged allegiance [to ISIS], in order to help the brothers and sisters of ISIS... and teach them how to make da'wa [preach] to people who have long lived in humiliation and do not know the laws of Allah. We are here, and we make da'wa to the children, to the elderly, to all the people."

Islamic State English-Language Magazine *Dabiq*: "Distribute Them Through All Forms Of Media, Including The Internet"

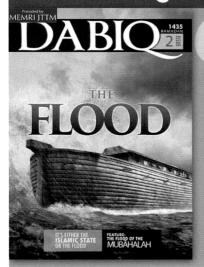

"*If you cannot perform hijrah for whatever extraordinary reason, then try in your location to organize bay'at [pledges of allegiance] to the Khalifah Ibrahim [IS leader Abu Bakr Al-Baghdadi]. Publicize them as much as possible. Gather people in the masajid, Islamic centers, and Islamic organizations... and make public announcements of bay'ah. Try to record these bay'at and then distribute them through all forms of media, including the Internet. It is necessary that bay'ah becomes so common to the average Muslim that he considers those holding back as grossly abnormal. This effort, insha'allah, will encourage Islamic groups to abandon their partisanship and also announce their bay'ah to the Khalifa Ibrahim. If you live in a police state that will arrest you over such bay'at, then use means of anonymity to convey your bay'ah to the world.*"

The Islamic State (ISIS) English-language magazine *Dabiq's* second issue urged followers to use "all forms of media, including the Internet" to promote pledges of loyalty to it.

"*Translation departments have a very important role in **delivering** the **message** to non-Arabic speaking people... Words in Arabic are not translated as they should be.*"

— *Haq wa Afsan Tafsir* magazine, June 2014

A pro-Islamic State (ISIS) jihadi English-language journal, *Haq wa Ahsan Tafsir*, is placing considerable emphasis not only on dissemination of its message but on how Islamic terms are presented and explained to potential recruits. It wrote in an article titled "Advice to the Media Translation Departments," about translating Arabic terms: "Translation departments have a very important role in delivering the message to non-Arabic speaking people... Words in Arabic are not translated as they should be." It continues "We Muslims must give the right name to all things and not follow shaytan [Satan]..."[122]

A jihadi who traveled to Syria, "Abu Turab," states on his Twitter account that he is 25, from Chicago, and has joined the Islamic State of Iraq and Syria (ISIS). In addition to Twitter, he is active on Facebook and Ask.fm. His first tweet, on December 21, 2013, was: "Internet, Restaurants, Cars, iPhones... Allah has made Hijra and Jihad in Sham [Syria] so easy... why are we still clinging to the Earth and hesitating?"

In another example, "Abu Al Bawd Irhabi" tweets an image and jokes, "ISIS & Jabhat Al Nusrah AQ SHAM guys caught working together in one internet room." On December 10, 2013, @JehadNews tweeted images of "one of the British mujahideen [fighting alongside] the Islamic State of Iraq and Al-Sham [ISIS]," who is shown using a Kindle with a gun on his lap.

British member of ISIS Using a Kindle.

A May 28, 2014 tweet by @jnob_isis showed how ISIS is sharing information hand to hand, and notes: "Ewessat (area southwest of Baghdad) / distribution of audio and visual publications by the Islamic State and Al-Sham."

On July 27, 2014, Jabhat Al-Nusra tweeted from its Urdu-language account a photo of mujahideen gathered around a laptop:

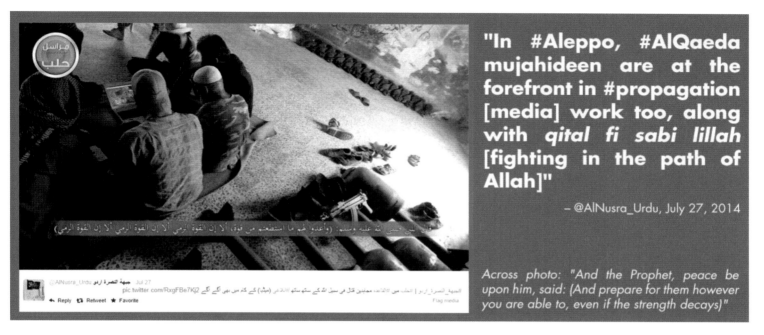

A July 29, 2014 tweet by @yo88_yo showed a laptop, smartphones, camera, and thumb drive, and requested followers' prayers:

A photo tweeted July 28, 2014 by @jondreporter drew attention to ISIS's use of a laptop, smartphones, tablet, thumb drive, and camera to disseminate media productions:

The cover photo of an official Islamic State Twitter account shows how the laptop is now a basic part of jihadi equipment:

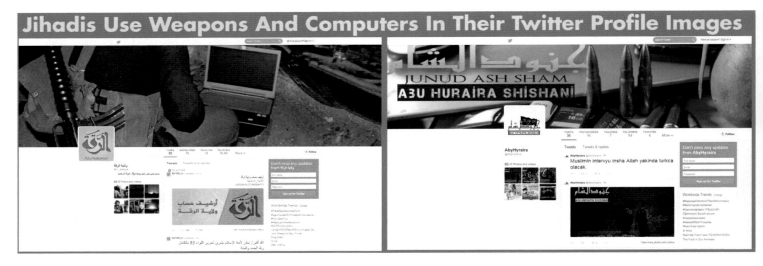

One of the scores of foreign jihadis in Syria tweeting about their experiences in jihad, including their daily activities, fighting, technological devices, and more, is a Danish jihadi calling himself Abu Fulan Al-Muhajir who as of this writing is fighting in Syria is tweeting his experiences in jihad. He began tweeting April 8, 2013, and as of this writing has posted over 2,200 tweets and has nearly 3,400 followers; his first tweets were about his travels to Syria via Turkey. He also tweets about his jihad experiences; for example, on September 1, 2013, he tweeted that he was "at tactical meeting with 3 other jihadi groups. 'Where should we attack?'..." and describes meeting jihadis from around the world, including from European countries such as Germany, Albania, The Netherlands, Sweden, Russia, Britain, Norway, and Denmark.[123]

Accessed October 3, 2013 and February 13, 2014, respectively.

Islamic State supporter ShamiWitness tweets prolifically, as a jihadi news aggregator who is given exclusive stories for distribution by various jihadi groups in Syria and Iraq. He bills his Twitter account as "Musings on bilad ash Sham, Sunni Revolutions, Economic Collapse, Post-industrial society, Technology, History etc. RTs [retweets] ≠ endorsements. Following ≠ endorsements." He is active on social media around the clock.

Twitter.com/Radicalislamist; Twitter.com/ShamiWitness.

Jihadi groups in Syria place a great deal of importance on online training. User Ansaar_Gbhat tweeted a photos of a young jihadi being trained and a computer monitor and wrote, "#Jabhat Al-Nusra establishes teaching and training courses for [graphic] design programs for the cubs of jihad in Al-Sham."

Another tweet showed the "technical electronic office" preparing "handheld communication pieces for the stationed brothers" in the field:

Jabhat Al-Nusra's English-language bulletin "Monthly Harvest," released via Twitter for download from the Internet Archive,[124] summarized its activities for Sha'ban (May-June) 2014. The magazine comprises mostly a collection of images; one shows a military leader "explaining the plans for storming the village of Umm Sharshooh (We Invade Them Now)" using a presentation via his computer.

In a tweeted image, Jabhat Al-Nusra leader Abu Maria Al-Qahtani of Al-Sharqiyya Operations gives a military training presentation on his computer to the Military Shura Council.

A group of British jihadis in Syria known as Rayat Al-Tawheed, which is active primarily on Facebook and Twitter but also on YouTube and Instagram, appears to function mainly as a graphic design team, creating jihad and martyrdom images for sharing via the jihadi forums and social media, with a focus on urging youths to join the jihad in Syria. In a video on the group's YouTube page, a jihadi in Syria speaking English with a British accent says, "You don't just come here and put on a tactical vest and grab a Kalashnikov and get a big beard and that's it. This is not just a thing you can put on Instagram or Facebook."[125]

The group tweeted on February 8, 2014 tweet that while Facebook, Tumblr, and YouTube had all been "shut down," "Twitter and insta[gram] the only ones keeping it real."

In a video, British Rayat Al-Tawhid member "Abu Abdullah" shows the living quarters of Western jihadis in Syria, commenting that they hardly constituted "five-star jihad" conditions. He pointed out a laptop and underlined the importance of the work done on it.

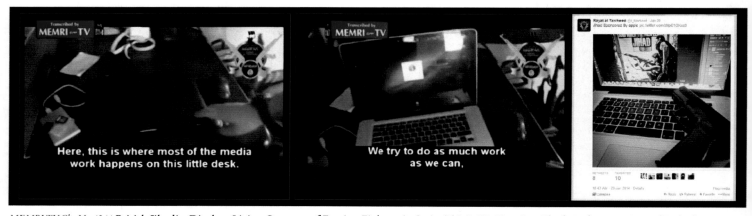

MEMRI TV Clip No. 4244, British Jihadist Displays Living Quarters of Foreign Fighters in Syria: This Is No Five-Star Jihad, April 20, 2014; Rayat Al-Tawheed tweet, January 20, 2014, accessed February 27, 2014.

Underlining their interest in technology and computers, jihadis in Syria also express interest in playing video games, for example with PlayStation, as Abu Turab tweeted on January 20, 2014 and as @Abuomer_kurdy tweeted on August 23, 2014:

Movie Night For Canadian Fighter In Syria: Orange Crush, Popcorn and an AQAP Video

Tweet by a self-identified Canadian in Syria, Muthanna Al-Kanadi @Muthanna88

Judging by their online activity, young jihadis from the West are highly familiar with and influenced by video games, especially ones simulating war and combat. Images circulated on their social network profiles include references to games such as "Call of Duty."

A British fighter who goes by the Twitter name Sayyad Al-Suri explicitly states that it was "Call of Duty" that inspired him to go abroad to join a jihad group. He wrote: "I grew up playing Call of Duty, so that game inspired me to do my duty." Jihadis view waging jihad as a personal obligation for every capable Muslim man.

MUJAHID AL-GHARBI TWEET - DECEMBER 25, 2013:

"I grew up playing Call of Duty, So that game inspired me to do my duty"

The photos below, from the Facebook pages of Abu Jibaal, an online supporter of Al-Qaeda and Jabhat Al-Nusra, and of Ahmadh Abdallah, also refer to the game.[126]

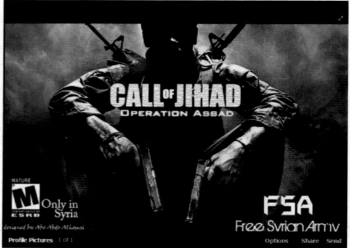

Calls To Engineers To Join The Islamic State July 2014 – By Islamic State (ISIS) Caliph Al-Baghdadi And Others, Including Foreign Jihadis

Historically, Al-Qaeda and its offshoots, including ISIS, have placed great importance on filling their ranks with experts in technology and computer science, as well as with engineers specializing in various fields. Spokesmen and members have issued calls to those with such skills to join them.

IS Leader Al-Baghdadi – The New "Caliph" – Issues A "Special Call" To Engineers To Join The Islamic State

Shortly after being declared caliph of the caliphate announced by the Islamic State (IS/ISIS), on July 1, 2014 Islamic State leader Abu Bakr Al-Baghdadi called on Muslims with needed skills and professions to join the Islamic State. He said: "We make a special call to... engineers of all different specializations and fields... their emigration is wajib 'ayn [an individual obligation]..."¹²⁷ Excerpts from Al-Baghdadi's statements were also published in Issue I of the Islamic State's English-language magazine *Dabiq*, released July 5, 2014.¹²⁸

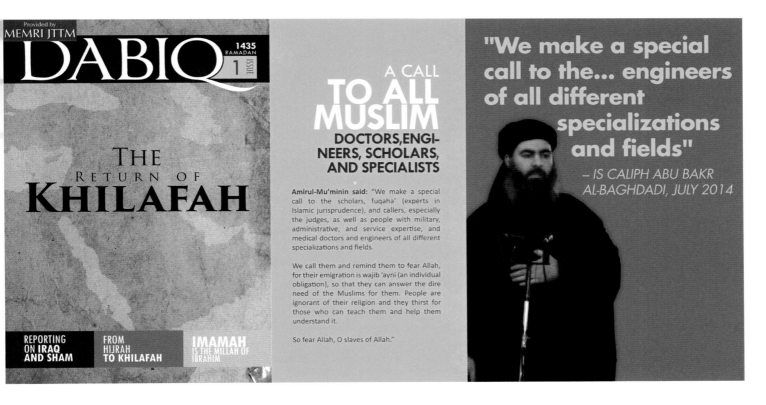

Canadian Jihadi In Syria: Come Fight In Syria – "We Need The Engineers" – And "If You Can't Fight... You Can Assist In Technology"

In a video released posthumously, on July 12, 2014, Canadian jihadi "Abu Muslim" calls on Muslims living in the West to come to Syria, and stresses: "You have to live as a full Muslim... so this [i.e. coming to Syria] means more than just fighting... We need the engineers, we need doctors, we need professionals, we need volunteers, we need fundraising, we need everything... There is a role for everybody. Every person can contribute something to the Islamic State... If you can't fight, then you can give money. And if you can't give money, then you can assist in technology, and if you can't assist in technology, you can use some other skills."¹²⁹

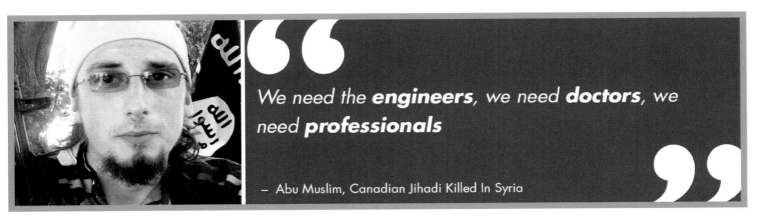

The Al-Qaeda affiliate Ansar Al-Islam, which announced on August 28, 2014 that it was disbanding and merging with ISIS,[130] was praised in a July 23, 2014 tweet by @Khorasan313: "Ansar al-Islam is known for their #Engineers & #Chemists who scared #US to extent that #US held a session to discuss about them in Congress."

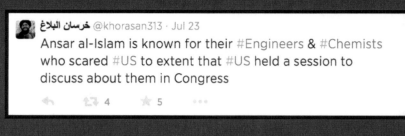

A May 7, 2014 Facebook announcement by "Democracy Is The Cancer, Islam Is The Answer" called for engineers to join ISIS: "If you want to join ISIS and don't want to fight as a combatant, you can instead contribute a lot if you are an engineer, doctor etc ISIS." The same message was circulated via Twitter, with a link to the Facebook announcement.

Twitter.com/ISIS_PAKISTAN, May 7, 2014.

Jihadi Hacktivist Groups Emerge In ISIS

Over the past two years, jihadi hacktivist groups have also emerged in Syria and Iraq. British hacker Junaid Hussein, aka Abu Hussain Al-Britani, from Birmingham, UK, travelled to Syria a year ago while under police supervision following his arrest for hacking emails of then-prime minister Tony Blair and posting his personal information online. He is now thought to be a main recruiter and mastermind of a widespread ring of young computer experts recruited by the Islamic State to hack into bank accounts of UK businesses and of celebrities, among them rapper P Diddy. He is believed to be teaching the new recruits how to crack codes and access sensitive information. A source said of the operation, "This is a new dawn of warfare. This is not a sporadic operation. The hackers are targeting the accounts of the rich and famous, VIP clients of banks and big businesses. This is an international fraud on an unprecedented scale and the result could be a bottomless pit of money to fund their campaign of terror."[131] At the time of this writing, security authorities believe that he is connected to the beheading of American photojournalist James Wright Foley.[132]

Twitter.com/AbuHussain102, accessed August 22, 2014.

In another example, a July 5, 2014 tweet announced: "The cyber mujahideen of the Islamic State have hacked the Iranian website iranefardamag.com."[133]

The Islamic Front (IF), an umbrella organization of a number of Islamist opposition groups in Syria, recently launched the "Electronic Islam Company," the IF's cyber warfare arm, which is tasked with hacking Syrian regime's websites and any pro-Bashar Al-Assad webpages.[134]

Tweets about the activity of the Electronic Islam Company first appeared on Twitter on May 2, 2014. Those tweets included links to a number of YouTube videos, which allegedly showed compromised webpages following the attacks. The videos, it should be noted, were all posted on the YouTube channel of Fursan Al-Islam ("Islam Knights").[135] It appears also, based on those videos, that the Electronic Islam Company belongs to Jaysh Al-Islam, which is one of the Islamist groups operating under the IF's name. The attacks also seem to be geared at the Facebook pages of various pro-Assad shabihas (thugs) and Hizbullah members. Kali Linux, a security-based OS and a hacker's go-to tool, appears to be used in the attacks. The number of members in the Electronic Islam Company is unknown, but it appears to be more than one individual based on the various computer backgrounds seen in those videos.[136]

IF's logo and the "Electronic Islam Company" on the back of computer (Source: zamanalwasl.net, May 7, 2014)

Death Photos, Eulogies For Jihadis Killed In Battle Used As A Recruitment Tool

Another use of U.S.-based social media is jihadis' dissemination of death photos and eulogies for jihadis killed in battle. For example, a eulogy posted on Facebook for Abu Qasura Al-Tunisi, a Tunisian who was an important contributor to the now-defunct Al-Falluja jihadi forum and who traveled to Syria to fight alongside the Al-Qaeda-affiliated Jabhat Al-Nusra, noted: "He joined Jihad with no help but Allah and Google Earth."[137] His death photo has also been disseminated and shared widely via Twitter by online jihadis – much as American kids share and collect baseball cards – and videos of the martyrdom operation in which he was killed, and of his funeral, can be viewed on YouTube. He apparently used Google Earth not only for joining jihad in Syria but also for other travels; posted on his Facebook page are photos of him in Western countries such as Italy before he went to Syria.

Left and center: Abu Qasura Al-Tunisi; right, Al-Tunisi, undated. Hanein.info/vb/showthread.php?t=296564, accessed December 24, 2013.

Skype – Fundraising And Media Interviews

Skype, the instant-messaging and VoIP service, has come into increasing use by Al-Qaeda and Al-Qaeda affiliates – to the extent that the Pakistani province of Sindh has moved to ban it and other smartphone apps such as Whatsapp (see below). Sindh Information Minister Sharieel Memon said that these networks are being used by terrorists who have switched to them from using regular mobile phone services.[138] It has even come to be used in communications between media and fighters in Syria; a February 3, 2013 *New York Times* report on Al-Qaeda's break with ISIS included statements from a Skype conversation with a rebel fighter in Idlib province.[139]

Mustafa Maya Amaya, the recruiter for European fighters in Syria, was very active online, operating several blogs and using social media platforms to recruit young men to jihad and to advertise his services as middleman between would-be recruits and the jihad organizations. He wrote on his blog: "Add me as a friend on Skype: I am part of a group... The normal doors [i.e. the regular recruitment channels] are closed, but not ours! There is a group of brothers that has left [for jihad]. I take care of assembling the brothers – sending them to their work place [i.e. jihad]. Now, I work on the Internet. I also send brothers to Syria, and there, yes, it's [i.e. the way in] open!!! I can work with you."[140]

In another example, Skype is being used by one jihadi group, the Al-Haq Brigade (part of the Syrian Islamic Front) to help recruit new trainees for its Al-Ansar Battalion training camp.[141] The Al-Haq brigade, which is part of the Syrian Islamic Front, provided Skype contact information to potential trainee recruits, inter alia on its Facebook page. The Facebook notice asked all those interested in joining the camp to apply to the battalion's representative via Skype. The notice adds that anyone interested in joining is requested to contact the representative of the Al-Ansar Battalion via Skype, using the ansar.camp account. It should be noted that Skype is commonly used by the rebels in Syria in general, and particularly by the jihadis there.[142]

Left: The notice on Facebook. Top, in image: Skype account information; below, in text, arrow: instructions for applying via the Skype account. Source: Facebook.com/LiwaAlhaq, June 30, 2013. Right: Facebook page of the Al-Ansar Battalion, with the same Skype address (center of image). Text: "Homs is calling you – come to jihad; Al-Ansar Training Camp." Source: Facebook.com/k.alansar, accessed July 8, 2013.

The Kuwaiti Twitter-based Ansar Al-Sham Campaign told potential donors to use both Skype and Whatsapp to contact campaign headquarters in order to contribute to its new "$5000 from Every House" campaign to provide military equipment for the mujahideen in Syria to "liberate the coastal area" of that country.[143]

Left: Fundraising campaign graphic with Skype and Whatsapp contact information. Right: Tweeted image under the hashtag "I will donate to the Mujahideen in Ramadan." Source: MEMRI Special Dispatch No. 5382, Fundraising Campaigns For Syrian Rebels Intensify During Ramadan,, July 23, 2013.

Skype contact information is also provided by the "Anshar Al-Sham" campaign that fundraises in Kuwait for the Al-Qaeda affiliate in Syria Jabhat Al-Nusra. Ansar Al-Sham's YouTube account features an appeal for donations which reads: "Oh you who want to do good! Your brothers are in a dire need for money. Don't withhold money from them. Support them with money and with the tongue... The statement lists further contact possibilities, including Skype, and Twitter.[144]

Screen shot from Ansar Al-Sham's YouTube clip, featuring the campaign's contact information against a backdrop of Jabhat Al-Nusra fighters

It was also recently reported that the Pakistan-based jihadi organization Lashkar-e-Taiba is using an exclusive Skype-type app thought to have been created by the Pakistani military's Inter-Services Intelligence to plan terror attacks.[145]

Whatsapp – Mobile Jihadi Messaging

The use of Skype is also promoted in tweets and retweets, as is the use of the mobile messaging app Whatsapp, for communication and fundraising purposes. The Texas-based WhatsApp, which in February 2014 was bought by Facebook, is now used widely by jihadis; for example, a senior member of the leading jihadi forum Shumoukh Al-Islam suggested that online jihadis use it to publish jihadi publications, interact on the Internet, and terrorize the enemies of Allah.[146] In addition, other terror groups such as Hizbullah are using it.[147] It is also being used to share photos.

Ttwitter.com/teletabiii

On September 15, 2014, Mufid Elfgeeh of Rochester, New York, a 30-year-old Yemeni-born U.S. citizen, was indicted for supporting a foreign terrorist organization and threatening to kill U.S. officers, following a federal investigation that began in early 2013.[148] According to U.S. Attorney William Hochul, Elfgeeh is accused of "providing material support" to ISIS; he put up cash to help send an unidentified person to Syria to fight for ISIS, helped that person get his passport, and bought handguns and silencers. The case began with tips to the FBI about Elfghee's anti-U.S. and pro-Al-Qaeda social media activity,[149] and a search warrant, obtained for three Whatsapp accounts associated with overseas telephone numbers, was based, inter alia, on subpoenaed records from Facebook and Whatsapp.[150] In the years leading up to his arrest, Elfgeeh, was very active on Twitter, with multiple accounts, two of which were cited in the application for his search warrant; in one of his accounts, he appeared to be close on Twitter to a prominent ISIS-supporting Twitter account.[151]

Photo: TWC News, September 18, 2014

On November 30, 2013, Danish jihadi in Syria Abu Fulan Al-Muhajir tweeted that he had texted "a fellow mujahid" on Whatsapp but had received a reply from another jihadi saying that the first man had been martyred.

 Abu Fulan al-Muhajir @Fulan2weet · Nov 30
Texting a fellow mujahid on WhatsApp, and getting a reply from another brother saying he has been martyred. Allahu akbar. May he be accepted

Expand   ← Reply   ⇄ Retweet   ★ Favorite   ••• More

Google Services – Mapping, Blogging, And Apps

Google Inc. technologies continue to be a favorite of jihadis through services such as Google+, which is now being used by Al-Qaeda and other jihadis to cultivate groups of individuals interested in joining jihad.[152] In addition, Google's Play Store, like Apple app stores, now sells jihadi content for Android and iOS phones and other devices;[153] Google Earth is used by jihadis like Abu Qasura Al-Tunisi; and the Google photo-sharing service Picasa is now being used by the Taliban to publish and share thousands of photos of U.S. troops in Afghanistan, including photos of wounded soldiers, of coffins of U.S. soldiers, and of U.S. tanks and other military vehicles attacked and damaged by the Taliban.[154] The Taliban also uses Picasa to release photos of "martyred" jihadis.[155]

Left, Riyah media foundation on Google+; right, Taliban on Picasa

A June 4, 2014 tweet of an image labelled "The plan to purge the Rawiyah area of YPG supporters" shows the use of Google Earth, and states: "Battle to liberate the village of Al-Rawiyah in north Raqqa Governorate from the hands of the YPG apostates, agents of the Nusayri regime." At the bottom of the screen, the user's Skype logo can be seen, as well as the Google Earth logo.

Twitter.com/raqqa98/status/474294312818470913/photo/1

A video from 2013 showed the Islamic Emirate of Afghanistan using Google Earth to plan an attack. Google Earth was used by the attackers in the 2008 Mumbai bombings,[156] and is also used by Urdu-speaking jihadis.[157] Also, a tweet by an ISIS commander used Google Earth to map areas for attacks.

ISIS commander using Google Earth, December 22, 2013; Islamic Emirate of Afghanistan planning an attack using Google Earth, 2013.

A pro-ISIS "news" app available for Android devices on Google Play, released in April 2014 and called "Fajr Al-Basha'ir" or "Dawn of Glad Tidings," is aimed, according to its description, at delivering "the latest news and events in Syria, Iraq, and the Muslim world." However, according to those behind its release and promotion, it actually "aims to support the Islamic State of Iraq and Syria by monitoring its entire field and media activities, and [by] publishing everything that relates to supporting it and defending it." The app has received a five-star rating from users.[158]

Instagram – Sharing Photos Of Al-Qaeda Leaders – And Used By Jihadis In Syria And Iraq

Instagram is an extremely popular online photo and video-sharing service established in 2010 that offers its users options for digitally enhancing and sharing their photos; users can follow other users' accounts and re-share images freely. The site boasts 90 million monthly active users, and says that its users post 40 million photos every day. Photos used on Instagram can be seen by visiting websites such as Statigram, a web viewer for Instagram that allows users to view and search photos.

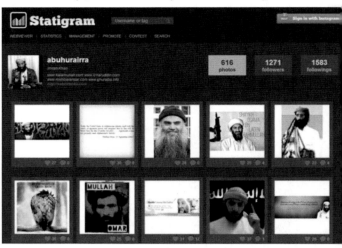

Since MEMRI first began monitoring Instagram, in 2012, there has been a tremendous increase of its use by online jihadis. A common theme of the photos that they post is images of and quotes by Al-Qaeda leaders such as Osama bin Laden, Anwar Al-Awlaki, Abu Mus'ab Al-Zarqawi, Ayman Al-Zawahiri, and many others. Another theme is the glorification of imprisoned jihadis: the Blind Sheikh Omar Abd Al-Rahman, Fort Hood shooter Maj. Nidal Hasan, underwear bomber Umar Farouk Abdulmutallab, and many others who have successfully attacked Americans.[159]

Instagram and its offshoots are used by jihadis in Syria. For example, Saudi teen Muadh Al-Jraish[160] is active on Instagram and on iPhoneogram, a web-based Instagram app.

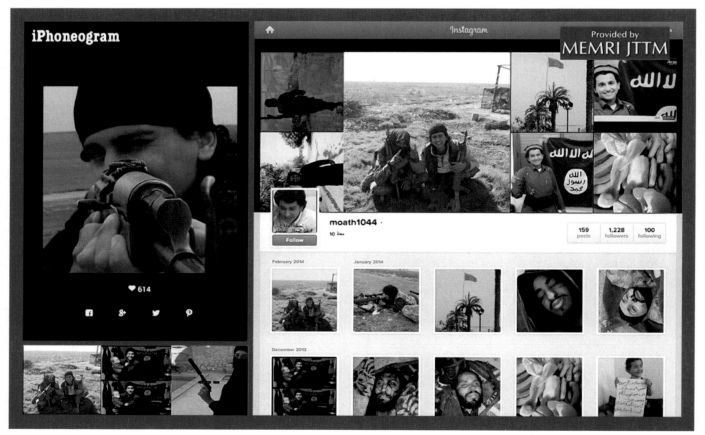

Top left: Al-Jraish's Iphoneogram (Instagram) cover photo, iphoneogram.com/p/678253772312080548_460753771/. Lower left: photos from his Instagram account; lower right image is a still from video of him firing a gun. Right: front page of his Instagram account.

Flickr – Snapshots Of Martyrdom

 Flickr, the popular online photo management and sharing application, is now one more of the large U.S. social media companies that are infested by jihadis and are being used to help drive jihad in Syria. For example, there are now Flickr pages that show foreign fighters martyred in Syria from countries including France, Sweden, and other European countries.[161]

Tumblr – Microblogging Jihad

 Another social medium used by jihadis in Syria is the microblogging platform Tumblr, which was bought by Yahoo in May 2013 for over $1 billion, and is rapidly gaining popularity among Western jihadis. A November 20, 2013 MEMRI report focused on one British jihadi's ongoing account of his journey to the jihad in Syria; his page also has a soundtrack with sounds of battle and men singing jihadi songs.[162]

Another jihadi using Tumblr was "Abu Layth," revealed to be Anil Khalil Raofi from Manchester, who studied engineering at Liverpool University.[163] He tweeted very actively about the Syrian jihad front.[164] On January 25, 2014 he wrote on

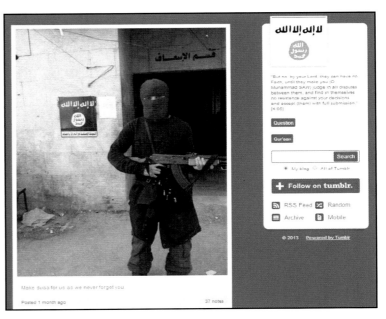

Tumblr, "not really planning on posting much, but yeah, it's available." By the next day, he had changed his mind about Tumblr's usefulness; he tweeted, "Actually tumblr is quite useful for long posts to be honest I'll probably post longer stories etc on there from now on and share the link." In the days that followed he posted some fairly lengthy items on the service. He was reported killed in northern Syria on February 15, 2014; as is often the case with foreign fighters in Syria who use social media, his social media accounts remain open.

Ask.fm – Jihadi Q&A; Kik – More Privacy

The Latvia-based Ask.fm is a social media website in which interactions take the form of questions and answers. It states on its website that it "allows for anonymous content which ask.fm does not monitor." MEMRI has published a number of reports[165] about jihadis in Syria using it to answer questions from readers wanting to know about joining the jihad there. Media have reported recently that it seems to have become a key communication tool for British-born terrorists fighting in Syria and their followers in the UK.[166]

For example, a British jihadi who goes by the alias of Abu Abdullah Al-Britani, and is believed to have travelled via Turkey to Syria to join ISIS and to be fighting close to the Syria-Iraq border, is active on Twitter, originally as @Al_Brittani, and, after that account was suspended, as @AlBrittani. However, Abu Abdullah is also active on Ask.fm, where he answers questions about jihad, about travelling to Syria, about weapons, fighting, and so on. On Ask.fm, he confirms that his old Twitter account was suspended, discusses how important the use of social media is to ISIS, refers to using the Canada-based smartphone messaging service Kik. He also, as of this writing, is active on Facebook, as Abu Abdullah Al-Britani, and on YouTube as Abu Abdullah.

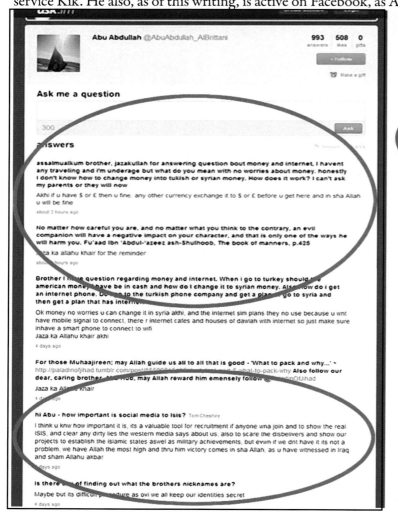

From Al-Qaeda To The Islamic State (ISIS), Jihadi Groups Engage in Cyber Jihad: Beginning With 1980s Promotion Of Use Of 'Electronic Technologies' Up To Today's Embrace Of Social Media To Attract A New Jihadi Generation

Other jihadis in Syria using Ask.fm have been asking readers to contact them via Kik, which has also been adopted by Western jihadi recruiters and foreign fighters in Syria to assist in recruitment and in transportation to and from Syria. For example, on his Ask.fm account, Abu Fulan Al-Mujahir discussed with a follower the use of Kik and another messaging program, Surespot, for purposes of jihadi communications:

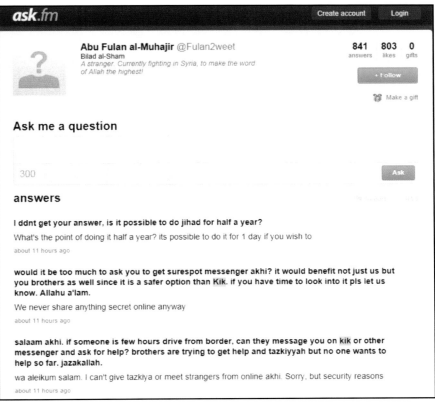

Abu Fulan Al-Muhajir on Kik. Webcache.googleusercontent.com/search?q=cache:PHdbg67Z414J:ask.fm/Fulan2weet+&cd=6&hl=en&ct=clnk&gl=us

Jihadis are also providing their Ask.fm and Kik contact information on their Twitter profiles.

"@MujahidAlShami" provides his Ask.FM and Kik contact information on his Twitter account (Twitter.com/MujahidAlShami). Left: "The Islamic State attacking Syrian Army Division 17 with 220mm mortars in Raqqah"; right: "There are more than 70 competent schools in #Raqqah #ISIS #Sham"

The late Abu Layth had an Ask.fm account in addition to his other social media accounts; he used it, inter alia, to give fellow UK citizens advice on how to join up, and to post a video in which he and other Brits openly discussed becoming suicide bombers.[167]

Twitter.com/r_tawheed1/status/488639944991457280

After its Twitter account was shut down, Rayat Al-Tawheed opened a subsequent account, @_tawheed1, and on July 14, 2014 used it to refer readers to its Kik account.

A foreign jihadi in Somalia, "Saahibul Hijratain," who appears to be from the U.K, is using Ask.fm, in addition to Twitter and Instagram. He wrote in December 2013 that foreign fighters still "continue to come to Somalia... a steady incoming flow" and added that "since all eyes are mostly focused on Syria atm, it has overshadowed Somalia and made the number of muhaajireen arriving seem smaller (well, not just Somalia but most Jihaad fronts)." On January 11, 2014, he wrote that his fight there was part of global jihad aimed at instating shari'a law not only in Somalia but throughout the world. In his words, after the liberation of Somalia, "we'll fix our gaze on Rome."[168]

Also, Saahibul Hijratain wrote on his Twitter account on September 27 that he had met with American Al-Shabab member Cabdulahi Ahmed Faraax, a.k.a Abdullah Al-Amriki, and expressed disbelief at the fact that the FBI is afraid of him: "Can't believe the FBI are scared of this guy LOL. V[ery] funny humble guy."

On December 14, he boasted of the jihadi groups' capabilities, and hinted at another 9/11-style terror attack to come. He wrote: "What you see about Jihad online is just a fraction of how amazing it is... And whatever you see from the Mujahideen online, be it their techniques, their advancement, their knowledge, their actions... That too is a fraction of the reality. Who knows, perhaps we have another 9/11 coming up :)"

Abu Turab, the Chicago jihadi in Syria, answered explicit questions on his Ask.fm account about how to get to Syria and instructed jihadis to message him for assistance crossing the border.[169] He conducts a conversation with a "Canadian journo" with whom, according to the conversation, he is also in contact on Twitter; he asks him for his credentials and gives him information about Canadian fighters there. He also gives information on other foreign fighters, noting which ones "got shahada [martyrdom]."[170]

"The Mujahideen before last night's battle" Saahibul Hijratain's Twitter account, January 15, 2014. (pic.twitter.com/qkQSjlKqOQ, Jan 15, 2014)

Abu Turab's Ask.fm profile image. Text accompanying reads: "American 'Jihadi' or wtever they are calling it..inshAllah may Allah allow more and more muslims to make hijra and jihad in his path!"

Jihadi Ask.fm user in Syria "Mohamedfezo" fields questions on his account from other Islamists and jihadis regarding which group has legitimacy in fighting in and establishing an Islamic state in Syria; the questions include requests for reasons and proof of his conclusions. In his responses, he references Al-Qaeda leader Ayman Al-Zawahiri and the late Islamic State of Iraq leader Abu Omar Al-Baghdadi; in some of his answers, he posts YouTube videos of Al-Baghdadi. He also posted a video of Al-Qaeda ideologue Abu Mus'ab Al-Suri.[171] Unlike other accounts of jihadis in Syria, this account includes no questions or answers regarding the jihad in Syria. He also provides his Twitter and Kik information on his Ask.fm home page.

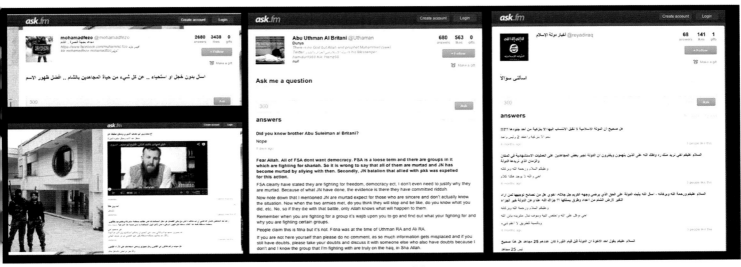

Ask.fm/mohamadfezo, with video of Abu Mus'ab Al-Suri, accessed February 15, 2014; Abu Uthman Al Britani's Ask.fm page; .

Yet another jihadi in Syria on Ask.fm, "Abu Uthman Al Britani" who says he is British/Bangladeshi, in his mid-20s, and with ISIS, is asked and answers questions about conditions fighting with ISIS, about how ISIS takes care of its fighters, about other foreign fighters, and more – and even fields numerous proposals of marriage. He notes that there are "loads of English-speaking people here," and that his love for martyrdom has grown and that he has "lived too long"; his profile image glorifies martyrdom. He refers readers asking questions such as how to reach Syria and whether he is interested in marrying a particular "sister" to his Kik account, responding to a reader's question, "why do our Mujahid brothers use Kik, is it safer or something?" with "Don't know, maybe because there's more privacy." and telling others, "Ask personal questions on kik."

AmreekiWitness, an Islamic State supporter who is prolifically active across popular social media, and who is one of many Americans and other Westerners who now vocally support ISIS on social media. He provides privacy information to his followers on his pages; for example, on his Twitter page on July 4, 2014, he admonished a reader who asked him about his recommendations "for a brother inside US who would like to fight against kufr": "Don't make these statements inside US unless you're operating through TOR and Ghost VPN." Asked on July 13, 2014 "Why are people asking about how to use TOR?", @AmreekiWitness replied, "To be anonymous online, they don't want the government seeing what they do and getting them in trouble." On both tweets, he links readers to the identical questions and answers on his Ask.fm account. AmreekiWitness' logo on his Twitter and Ask.fm accounts, uses the seal of the U.S. State Department, and his profile says that he is "Dedicated to raising awareness about the upcoming conquest of the Americas, and the benefits it has upon the American people." His Twitter account links also to his Wordpress.com blog.

The Ask.fm accounts highlighted in this report are a mere sampling, and do not constitute an exhaustive list. There are hundreds, if not thousands, of accounts, and more are appearing every day.

SoundCloud – Jihadi Recordings

The most recent social media to be adopted by jihadis is SoundCloud, the Berlin-based social networking platform that is now the world's second biggest streaming music service,[172] that allows users to upload and share audio content for free – a kind of YouTube for audio.[173] It is now being used by jihadis to share content ranging from Al-Qaeda messages and sermons by bin Laden and Al-Awlaki to ISIS nasheeds (religious songs) and even ISIS radio stations.

For example, the UmTwinz account has compiled a playlist of 23 tracks about the late American jihadist cleric Anwar Al-Awlaki, each of which has received an average of 200 to 300 plays. There is also a large amount of ISIS content on SoundCloud.

From Al-Qaeda To The Islamic State (ISIS), Jihadi Groups Engage in Cyber Jihad: Beginning With 1980s Promotion Of Use Of 'Electronic Technologies' Up To Today's Embrace Of Social Media To Attract A New Jihadi Generation

Additionally, content on SoundCloud is being disseminated via other social media. For example, the AQAP media wing Al-Malahem tweets files from the ISIS radio station Al-Bayan, in Ninawa, Iraq, which broadcasts news summaries and latest developments, and voice recordings of commanders are also disseminated.

ISIS's Extensive Use Of Social Media

On August 29, 2014, the Islamic State (ISIS) released the third issue of its English-language magazine *Dabiq*, which is aimed primarily at Western Muslims. In its foreword, this issue of the magazine, which was titled "A Call To Hijrah," listed a series of "related events that the Obama administration and western media tried to ignore when discussing the [U.S. air]strikes and the consequential execution of James Foley." The seventh one read: "Upon receiving the threat and prior to the execution, Obama scurried to prevent knowledge of the affair from reaching his citizenry. His administration immediately ordered a number of online social networks to shut down all Islamic State media accounts, including accounts of Islamic State supporters."[174]

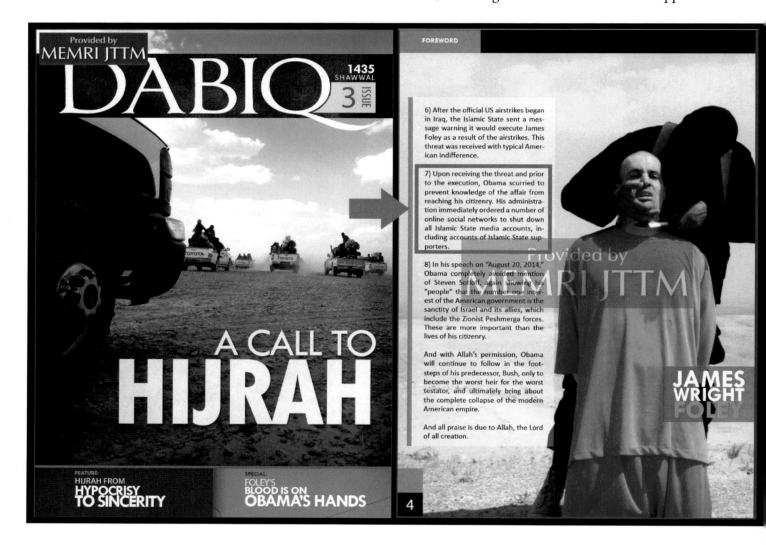

Ahrar Al-Sham Leader Hassan Abboud: "It's Very Strange How [ISIS] Has Been Able To Advertise Gory Executions And Beheadings Against The Social Media Websites Rules"

Highlighting the fact that ISIS has put a tremendous effort into their social media activity, and is garnering more and more headlines in Western media, other jihadi organizations are expressing some dissatisfaction online over this situation. For example, on August 20 and 21, 2014, Hassan Abboud, the leader of Ahrar Al-Sham and head of the political bureau of the Islamic Front alliance, tweeted criticism of Islamic State after it beheaded American photojournalist James Wright Foley on August 19, 2014 and disseminated the video of the execution via Twitter. Abboud wrote that "IS relies heavily on social media in order to spoil the image of Islam, attract new devotees, and lure young people into their ideology"; the previous day, he wrote, "It's very strange how the IS has been able to advertise gory executions and beheadings against the social media websites rules. #Foley." Abboud was killed in the September 10, 2014 bombing of Ahrar Al-Sham headquarters in Idlib province, northwest Syria.

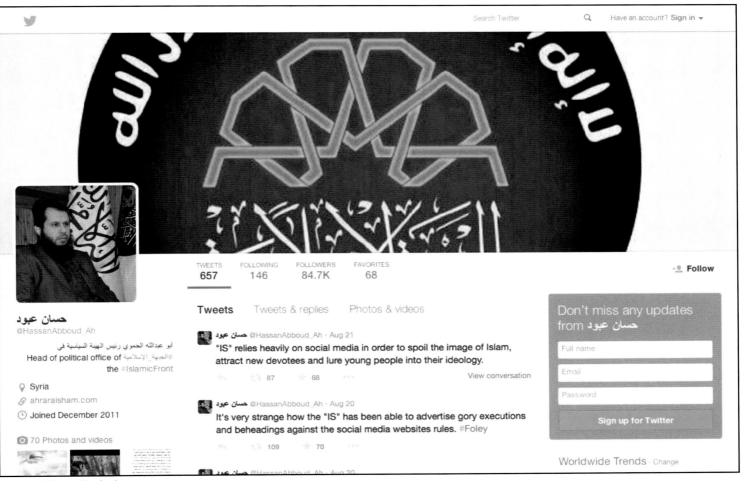

Twitter.com/HassanAbboud_Ah

ISIS "Media Points" Campaign Provides Religious And ISIS Information To The Public

In addition to its social media efforts directed towards the West and the rest of the world, jihadis are also focusing on areas already under their control. Barham Salih, former deputy prime minister of the Kurdistan Regional Government of Iraq, discussed the importance of the Internet in ISIS-held Mosul in an interview with Charlie Rose aired on October 21, 2014. He said that "a telecommunication company from where I come from is running – has been running – an Internet service in the city of Mosul. ISIS has taken control of that place, has appointed, apparently, [a] Western-educated person to be in charge of Internet service in that town. And they talk to this Internet provider over the Internet about how that service should improve or not improve, what to do."[175]

As reported on July 24 and August 7, 2014 by the MEMRI JTTM, ISIS's "Media Points" information campaign in Syria and in Iraq comprises information booths scattered throughout areas controlled by ISIS and provides religious and ISIS-related materials to the public. The campaign, which ISIS launched during the Eid Al-Fitr holiday, in late July 2014, consists, according to the organization, of "stationary spots [that are] scattered geographically across general areas" that distribute da'wa-related materials, screen "jihadi productions on display screens," and "transfer [those materials] to mobile [devices] via

Bluetooth [technology] and RAMs..." The campaign has generated much interest among the locals and has been received very well, according to ISIS.[176]

On some booths, banners show how the content they offer can be accessed: audio, video, Facebook, RSS, and more. They also serve as a venue for instructional gatherings and for screenings of ISIS videos, such as the videos of the execution of James Foley and Steven Sotloff.

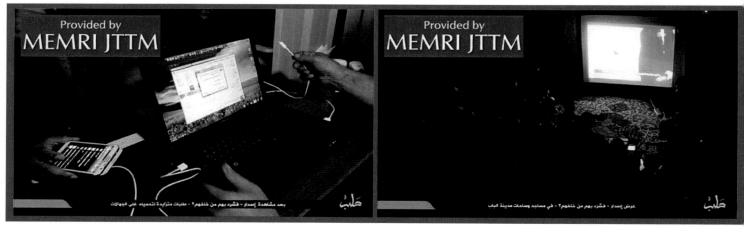

ISIS stated that the campaign has been well-received; the following are images from Mosul, Iraq:

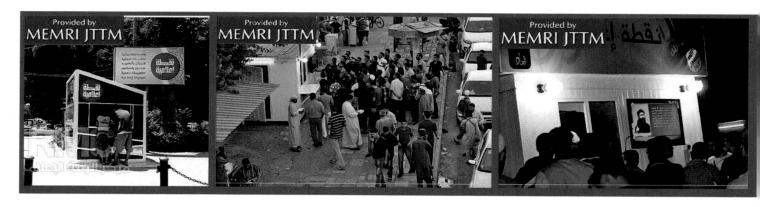

ISIS has also expanded its Media Points campaign to hospitals; a series of tweets by the AQAP media arm Al-Malahem show laptops set up in a hospital in Raqqa province in Syria, with photos reading "Media Spot in one of the hospitals of Islamic State in Raqqa province."

On November 5, 2014, ISIS released its first video about its Media Points campaign, focusing on its new Media Points booths in Mayadeen, Syria. The video featured local residents, including children, crowding around the booth and receiving information.

Also, in a video uploaded to YouTube on July 4, 2014, an ISIS leader speaks to a group of people in the Iraqi town of Anah, Anbar province, asking all those with weapons to surrender them to ISIS in return for their safety. He says that Islamic State Caliph Abu Bakr Al-Baghdadi has given orders for repentance to be accepted – presumably the repentance of Sunnis who were fighting ISIS. Members of his audience capture his statements with their smartphones.

An Islamic State Leader Gives Statements In Iraqi Village; Villagers Use Smartphones To Record Him

Friendica And Diaspora

On July 12, 2014, it was announced via Twitter that ISIS has decided to move its media companies Al-I'tisam,[177] Ajnad,[178] and Al-Hayat[179] to the France-based social media service Friendica.eu instead of Twitter, because of the constant shutdown of these Twitter accounts.

The Friendica Project bills itself as "a worldwide consortium of software developers creating decentralized social platforms and technology for the coming post-Facebook world.... quietly working behind the scenes to provide the most reliable, full-featured, and extensible alternative to the monolithic providers."[180] It allows users to create their own social media server and also to integrate news feeds from other platforms such as Twitter and Facebook without going through them. It is much more private, as it circumvents social networks' collection of data about users, and is decentralized, going through different servers.

Twitter.com/Battar1_IS/status/488012518498242560

diaspora*

Following Twitter's August 2014 shutdown campaign against ISIS accounts, and particularly following the August 19, 2014 ISIS beheading of American journalist James Wright Foley, ISIS accounts were opened on the Diaspora social network.

On August 19, 2014, information on all ISIS accounts on Diaspora was posted in Justpaste.it, along with a tutorial on joining and using the service.

Joindiaspora.com/u/nynwa_news, accessed August 19, 2014; justpaste.it/Wilayat1, accessed August 19, 2014.

VK.com

 As Twitter has been shutting down accounts, Vkontakte, or VK.com – the world's second-biggest social network after Facebook and used by Russian-speakers worldwide – is being increasingly used by jihadis; posts on other social media accounts are keeping followers up to date on the changeover to the service. Notably, the third issue of the Islamic State's English-language magazine *Dabiq* was released via the VK.com account of the ISIS media company Al-Hayat, on August 29, 2014.[181] Until recently, the foreign fighters in Syria using it were generally from Russia, mostly from the North Caucasus republics.

About a week after the Islamic State began using the service, VK.com shut down their accounts.

List of new Islamic State VK.com accounts, disseminated on Tumblr (source: Daawla.tumblr.com/post/96709116088); IS tweets on VK.co...

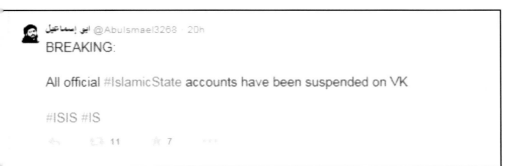

Webcache.googleusercontent.com/search?q=cache:i3X60IE2To0J:https://twitter.com/AbuIsmael3268+&cd=1&hl=en&ct=clnk&gl=us

In another example, this account, whose owner identifies themselves as Chicago-based and as a Chicago Theological Seminary alumnus, is now closed.

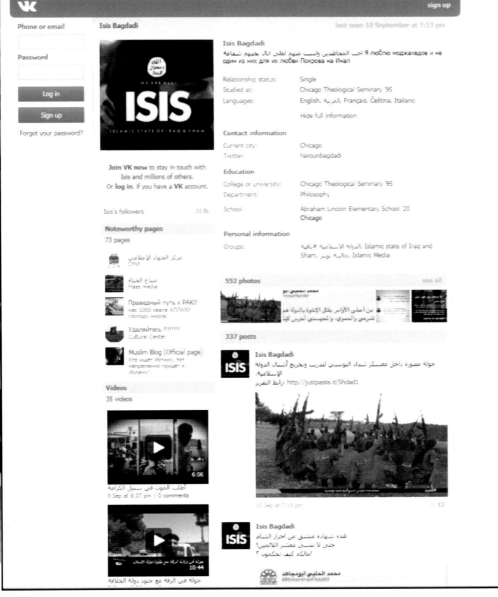

Vk.com/id267161804

JustPaste.it

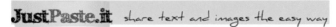 The Poland-based European JustPaste.it service is being used by ISIS to upload images of executions, beheadings, and massacres, as well as for lectures and other media productions. It is also being used by the organization to keep followers and sympathizers informed of new Twitter and other social network accounts. Other jihadi organizations are using it to publish and disseminate jihadi content, including reports on suicide attacks and other jihad-related news stories, official statements by jihadist commanders, and other material aimed at radicalizing Muslim youth.[182]

For example, on June 12, 2014, Australian Islamist and ISIS supporter Musa Cerantonio posted on his Facebook page a Justpaste.it link to a speech by ISIS spokesman Abu Muhammad Al-'Adnani; on August 27, 2014 @Wilayat_IS tweeted links to content on Justpaste.it, and on August 24, 2014 an ISIS Twitter account directed followers to Justpaste.it:

Algorithms: Helping Jihadis Find Each Other – Even After Social Media Accounts Are Shut Down

As this report has highlighted, Twitter and other social media platforms were finally moved to shut down accounts belonging to ISIS and other jihadi groups following the beheadings of Westerners by ISIS. Twitter CEO Dick Costolo even tweeted on August 20, 2014, "We have been and are actively suspending accounts as we discover them related to this graphic imagery. Thank you."

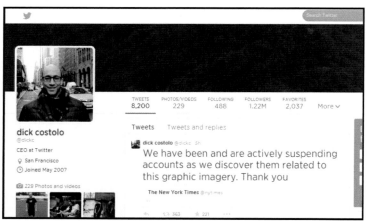 

However, as media interest in these stories waned, groups whose Twitter accounts have been shut down have reappeared there, quickly regaining their loyal followers for several reasons. New accounts immediately gain followers, through algorithm-generated recommendations to users based on their history. For example, Twitter itself provided the followers of the old accounts with the information on the new IS accounts and recommended that they follow them. Twitter explains that its algorithm "determines highly personalized suggestions based on the accounts you currently follow, the accounts followed by the users you have chosen to follow, and how other users express interest in Tweets sent from these accounts." It also says that it uses information that it collects on users' browsing habits on websites that contain the Twitter "share" buttons in order to recommend to its users whom to follow.[183]

JIHADIS TWEET THEIR REACTIONS TO TWITTER'S SUSPENSION OF ISLAMIC STATE ACCOUNTS

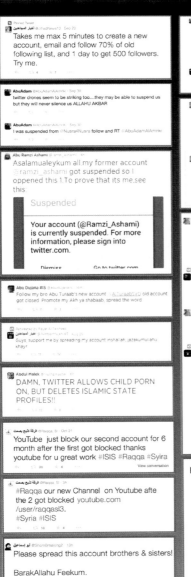

VII. The Future of Online Jihad – The Coming Battle With The Cyber Army Of Al-Qaeda And Its Offshoots

Twitter.com/AlTuraabSVD

This generation's activists of Al-Qaeda and its offshoots are younger and Internet savvy. They have heeded previous Al-Qaeda leaders' calls to turn to the Internet. They are interconnected, via Facebook, Twitter, YouTube, Instagram, Flickr, and every other new social media platform, adopting them almost as soon as they are created – just like the younger generation in the West. Like their Western counterparts, they have smartphones, tablets, and other devices; they have accounts with cellular carriers including contracts and monthly bills, and, like everyone else, they purchase and use apps.

With the combination of social media and mobile devices, jihadi outlets can make sure that their content is viewable anywhere, anytime, as shown by the tweet below of ISIS member viewing talks by Australian Islamist and prolific ISIS supporter and promoter Musa Cerantonio on their smartphones.

Jihadis have been quick to embrace the concept of using YouTube and other video sharing services, taking advantage of this technology to provide courses and training in explosives manufacture, weapons training, and hacking. Today, anyone can receive tweets or Facebook posts from Al-Qaeda and its offshoots, and other terrorists groups, directly to their cellphones, in real time – and can immediately share them far and wide.

Left: Smartphone jihad. Twitter.com/Hamza_99. Right: "Supporters of Jabhat Al-Nusra" tweets "An explanation of the plan [i.e. suicide bombing, carried out] by the martyr Abu Abd Al-Rahman Al-Shami, may Allah accept him. Aleppo." Twitter.com/Ansaar_Gbhat/ status/462369352395014144/photo/1

From Al-Qaeda To The Islamic State (ISIS), Jihadi Groups Engage in Cyber Jihad: Beginning With 1980s Promotion Of Use Of 'Electronic Technologies' Up To Today's Embrace Of Social Media To Attract A New Jihadi Generation

Essential jihadi equipment. Twitter.com/omarz7/status/463677205491359744/photo/1/large

These young jihadis are also experimenting with cyber jihad and with methods such as hacking and other techniques. It can be expected in the future that Al-Qaeda's cyber army's activities will become a daily reality; in the wake of what is happening in Syria in this regard, we can anticipate that every major conflict in the future involving jihadis and Al-Qaeda will follow the same pattern and involve extensive use of social media and cyber activity.

All this poses a direct threat to the West. Western countries are becoming concerned about their nationals returning home after fighting in Syria – where they are developing extensive networking skills amongst themselves using social media on their smartphones and other devices. There have already been arrests for cyber jihad in the past few years, in the U.S., the U.K., Canada, Spain, Belgium, Denmark, Norway, France, Italy, Switzerland, Russia, and other Western countries, on charges including making terror threats via Facebook,[184] posting terror-inciting material to websites and blogs,[185] inciting terrorism via YouTube,[186] and downloading jihadi media from the Internet.[187] The sentences for those convicted have ranged from significant fines to up to 20 years in prison. Authorities will have to learn new ways of tracking, following, and monitoring these developing jihadi communication networks.

Right now, as highlighted in this report, Al-Qaeda and its offshoots are actively seeking out and calling on individuals with technical and engineering skills – for hacking and downing drones, for hacking into the "private encrypted network of the Pentagon" (because "the security check in Muslim countries... is relatively low as compared to the Western Kafir countries...")[188] and for those who "want to join ISIS and don't want to fight as a combatant" who can "instead contribute a lot if you are an engineer, doctor, etc" and who "scare the U.S."[189]

Urging engineers and others with in-demand skills to join the jihad and fulfill their role as a Muslim is becoming increasingly common from all levels – from Islamic State leader and Caliph Abu Bakr Al-Baghdadi, who issued "a special call to... engineers of all different specializations and fields... their emigration is wajib 'ayn [an individual obligation]..."[190] to the rank and file, such as Canadian jihadi "Abu Muslim" in Syria "You have to live as a full Muslim... so this [i.e. coming to Syria] means more than just fighting... We need the engineers, we need doctors, we need professionals..."[191]

In light of its growing awareness of online security and vulnerability in the post-Snowden era, Al-Qaeda and its subsidiaries and offshoots are seeking more secure ways to use the Internet for their purposes. For example, in its most recent upgrade, on July 12, 2014, the Al-Qaeda-affiliated Global Islamic Media Front (GIMF) released an updated version of its Android mobile encryption software, citing as a reason for doing so the fact that "global [communications] companies" are now cooperating with "international intelligence agencies."

From Al-Qaeda To The Islamic State (ISIS), Jihadi Groups Engage in Cyber Jihad: Beginning With 1980s Promotion Of Use Of 'Electronic Technologies' Up To Today's Embrace Of Social Media To Attract A New Jihadi Generation

On September 26, 2014, online ISIS supporters launched a social media campaign aimed at stopping the organization's supporters from posting any information it that could be used by its enemies to target its fighters, and from leaking organization statements before their official release. The campaign was circulated on Twitter using the Arabic hashtag #Media_Blackout_Campaign.

One document shared in the campaign contained instructions on disabling the geolocation function on mobile devices and on computers, so that their users cannot be pinpointed and possibly targeted. The following are images from the instructions:[192]

"Your device, phone, and browser, reveal your location...how do they pinpoint your location? And how [can you] disguise yourself?"

Jihadi groups are also beginning to use Bitcoin, the crypto-currency widely used in the Dark Web – the part of the Internet that cannot be accessed directly and is not indexed by common search engines – to support jihad and the mujahideen. This, according to a recommendation in an article titled "Bitcoin Wa' Sadaqat Al-Jihad" – or "Bitcoin and the Charity of Jihad" – published by the English-language ISIS-supporting blog alkhilafaharidat.wordpress.com, will allow jihadis to use Bitcoin, which it says is both shari'a compliant and untraceable by "kafir governments," to easily "send millions of dollars' worth of Bitcoin instantly from the United States, United Kingdom, South Africa, Ghana, Malaysia, Sri Lanka, or wherever else right to the pockets of the mujahideen."[193]

Al-Qaeda and its affiliates and offshoots are increasingly relying on the Dark Web to conceal their activity. It has long been perceived as immune to law enforcement surveillance, and allows jihadis to communicate security, to operate hidden jihadi webpages and forums, to funnel money to jihad, and more.[194]

On September 25, 2014, ISIS supporters circulated an instruction sheet that included various recommendations on maintaining online anonymity. The sheet, which was called "Ways of Hiding from the Crusader Alliance," was written by Abu Khadija Al-Muhajir, aka Iraqe Hacker. Al-Muhajir wrote: "After the declaration of the abhorrent Crusader alliance against the Islamic State and its supporters, it became mandatory upon us to be careful of them since the targeted weren't only the mujahideen, but their supporters as well. And targeting the mujahideen will be [done] by bombing their locations. As for us, the supporters, we don't rule out that our homes will be targeted [as well] after our locations are determined, [and that will be done] either by bombardment or by arrest."[195]

On August 27, 2014, Al-Aan TV reported on a laptop belonging to a member of ISIS, a Tunisian national named Muhammad, that it said was captured by "moderate rebels in Syria." According to the report, the laptop had on it, in addition to many jihadi speeches and nasheeds, thousands of jihadi documents that its owner had published, mostly on the Dark Web. One of them was a detailed 19-page document about making biological weapons and about spreading " chemical or biological agents in a way to impact the biggest number of people."[196]

 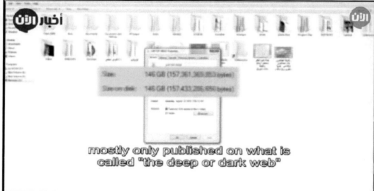

Since the publication of the 9/11 Commission Report in July 2004, counterterrorism officials have been warning about the threat of cyber jihad – but as of yet, little has been done. Over eight years ago, the 2006 National Strategy for Combating Terrorism stated clearly: "Increasingly sophisticated use of the Internet and media has enabled our terrorist enemies to communicate, recruit, train, rally support, proselytize, and spread their propaganda without risking personal contact... Our enemies... use today's technologies with increasing acumen and sophistication. This is especially true with the Internet, which they exploit to create and disseminate propaganda, recruit new members, raise funds and other material resources, provide instruction on weapons and tactics, and plan operations. Without a communications ability, terrorist groups cannot effectively organize operations, execute attacks, or spread their ideology... We will seek ultimately to deny the Internet to the terrorists as an effective safehaven for their propaganda, proselytizing, recruitment, fundraising, training, and operational planning."[197]

Over three years ago, the 2011 National Strategy for Counterterrorism warned: "Global communications and connectivity place Al-Qaeda's calls for violence and instructions for carrying it out within easy reach of millions." It went on to stress, "Be it in traditional media or cyberspace, a successful U.S. strategy in these domains will focus on undermining and inhibiting Al-Qaeda's ideology while also diminishing those specific factors that make it appealing as a catalyst and justification for violence."[198]

A photo making the rounds on Twitter shows Al-Qaeda leader Ayman Al-Zawahiri's envoy in Syria Abu Khaled Al-Suri, who was killed there, with his smartphone (source: Abo_al_hassan/status/438502810242789376/photo/1/large, accessed February 27, 2014).

It has been almost a decade since the U.S. government first stated that it would deny terrorists use of the Internet, but this has not yet come about; in the meantime, Al-Qaeda's online efforts have exploded. As esteemed U.S. Naval Postgraduate School professor John Arquilla wrote in his December 12, 2009 article "How to Lose a Cyber War" in *Foreign Policy*: "U.S. President Barack Obama often speaks about his central strategic objective of denying Al-Qaeda its haven in Waziristan, but he says nary a word about taking away its 'virtual haven' in cyberspace. This omission is more than his alone, as none of the key military, intelligence, and law-enforcement arms of the U.S. government have done much to curtail terrorist use of the Net..."[199]

While some Western governments are taking action, inter alia by closing down terrorist websites and social media accounts, high tech and social media companies – the beating heart of today's cyber jihad – must fulfill their obligation to make it difficult, if not impossible, for Al-Qaeda and other designated terror groups and their followers to use the Internet to propagate their ideology, recruit members and followers, fundraise for jihad, and other purposes highlighted in this report.

The current use of the primarily U.S.-based social media by Al-Qaeda and other jihadi groups to incite to jihad and spread their ideology, recruit to their cause, and fundraise has Western leaders extremely concerned. As Professor Arquilla stated, it is time for the Obama administration to take action. One of President Obama's first steps should be to bring together all leaders in the technology field to formulate strategies to address this growing problem.

All heads of social media and other tech companies whose services are used by Al-Qaeda and other jihadi and terrorist entities should be called to testify on Capitol Hill and to explain what they doing – and what they are not doing – to deal with it.

This is exactly what the U.K. House of Commons did; in early 2013, its Home Affairs Committee extensively questioned executives from Twitter, Facebook, and YouTube regarding jihadi content on their respective platforms.[200] Additionally, in late April 2014, the *Daily Mail* (UK) published an article titled "Spy Chiefs Warn PM: Internet Giants Including Google and Facebook Are Shielding Terrorists and Pedophiles"; it said that PM David Cameron had been warned that these companies are undermining UK national security.[201] Also, on September 15, 2014, France's parliament began debate on a new anti-terrorism bill that would give authorities power to block websites that defend or encourage terrorism and to require Internet service providers (ISPs) and website hosts to remove any such content.[202]

On October 8, 2014, the European Commission summoned major U.S. technology companies, including Facebook, Google, Twitter, and Microsoft, to a "private" meeting in Luxembourg on terrorist use of the Internet, against the background of the "flow of so-called foreign fighters" as well as "calls for electronic jihad that the EU is facing." The meeting was with ministers

From Al-Qaeda To The Islamic State (ISIS), Jihadi Groups Engage in Cyber Jihad: Beginning With 1980s Promotion Of Use Of 'Electronic Technologies' Up To Today's Embrace Of Social Media To Attract A New Jihadi Generation

from all 28 EU member states and members of the Commission. The European Commission stated about the meeting, "There is strong interest from the European Union and the ministers of interior to enhance the dialogue with major companies from the Internet industry on issues of mutual concerns related to online radicalization."[203]

Underlining the threat posed by cyber jihad, CIA director John Brennan said, on September 18, 2014 at the Intelligence and National Security Summit, that the "murderous, barbaric, criminal gang" that is ISIS is "something that has to be cauterized immediately" and that it will "be with us for a generation." Underlining that "we can't kill our way out of this," he added that ISIS must be destroyed, "not only from a standpoint of their ability to carry the attacks but the drivers that are sustaining them."

Those "drivers" are the same ones that are propelling not only ISIS but also Al-Qaeda and other jihadi organizations forward. Hundreds of thousands of young Muslims are now following them online; while not all of them are necessarily supporting these groups, these followers constitute a very large pool of potential recruits. Some Western countries are taking action against this online activity, but Capitol Hill and the U.S. administration are not following their example. As the U.S. government and U.S. social media companies continue to do nothing, an entire generation is continuing to be radicalized online.

Steven Stalinsky is Executive Director of MEMRI; R. Sosnow is Head Editor at MEMRI.

VIII. Endnotes

1. *Al-Quds Al-Arabi* (London), February 9, 2002. In addition, former U.S. Marines intelligence officer and homeland and national security expert Dan Verton predicted the current state of terrorist activity online in his 2003 book Black Ice: the Invisible Threat of Cyber-Terrorism.

2. *Sun-Sentinel*, August 21 2005.

3. Abd Al-Bari Atwan, 2008, "The Secret History Of Al-Qaeda," Chapter 4.

4. Abd Al-Bari Atwan, 2008, "The Secret History Of Al-Qaeda," Chapter 4.

5. *San Francisco Chronicle*, September 21, 2001.

6. Cryptome.org/usa-v-ubl-08.htm, February 21, 2001, accessed May 1, 2014.

7. MEMRI Inquiry & Analysis No. 886, The Life And Legacy Of American Al-Qaeda Online Jihad Pioneer Samir Khan – Editor Of Al-Qaeda Magazine 'Inspire' And A Driving Force Behind Al-Qaeda's Push For 'Lone-Wolf' Terror Attacks In West, September 27, 2012.

8. Abd Al-Bari Atwan, 2008, "The Secret History Of Al-Qaeda," Chapter 4.

9. WCVB.com, October 15, 2014.

10. *New York Post,* September 13, 2014

11. fbi.gov/newhaven/press-releases/2013/two-british-nationals-plead-guilty-to-terrorism-related-charges-in-new-haven-federal-court, December 10, 2013.

12. *Sun-Sentinel*, August 21 2005.

13. BBC.co.uk, February 15, 2002.

14. *Wall Street Journal*, November 11, 2002.

15. Abd Al-Bari Atwan, 2008, "The Secret History Of Al-Qaeda," Chapter 4. For more on statements by bin Laden, Al-Zawahiri, and Al-Mujahidoun, see MEMRI Special Dispatch No. 1826, Major Jihadi Cleric and Author of Al-Qaeda's Shari'a Guide to Jihad Sayyed Imam vs. Al-Qaeda (Part II): Al-Zawahiri Was Sudanese Agent – Sudan's VP Ali Othman Taha Hired Him to Attack Egypt; Ban on Jihad against Egyptian Regime in Egypt; Summary of Imam's New Right Guidance for Jihad Book, March 18, 2008.

16. MEMRI Inquiry & Analysis No. 827, Ansar Al-Mujahideen English Forum (AMEF) The Main English Language Forum for Al-Qaeda and Its Western Followers: Information and Communication Technology Thread Offers A Virtual Training Center for Online Jihad and Cyber Warfare Including Weapons Training, Hacking & Encryption, and Lessons in Becoming a Suicide Bomber, April 20, 2012.

17. See MEMRI Special Dispatch No. 3990, Al-Qaeda Announces Launch of New Jihadi Website, July 11, 2011.

18. See MEMRI Special Dispatch No. 1039, Terrorist and Inciting Messages on Yahoo, December 6, 2005.

19. See MEMRI Special Report No. 31, Islamist Websites and Their Hosts Part I: Islamist Terror Organizations, July 16, 2004; MEMRI Special Report No. 35: Islamist Websites and their Hosts Part II: Clerics, November 11, 2004.

20. See also MEMRI JTTM Report No. 374, The Enemy Within: Where Are the Islamist/Jihadist Websites Hosted, and What Can Be Done About It? July 19, 2007.

21. See MEMRI JTTM report Is Shutting Down Cyber Jihad Possible? The Real Scope of Cyber Jihad, December 7, 2009.

22. In late 2013 and early 2014, the AMEF website has been mostly down, but the URL remains active.

23. See MEMRI Inquiry & Analysis No. 827, Ansar Al-Mujahideen English Forum (AMEF) - The Main English Language Forum for Al-Qaeda and Its Western Followers: Information and Communication Technology Thread Offers A Virtual Training Center for Online Jihad and Cyber Warfare Including Weapons Training, Hacking & Encryption, and Lessons in Becoming a Suicide Bomber, **April 20, 2012.**

24. See MEMRI JTTM report Cyberspace as a Combat Zone: The Phenomenon of Electronic Jihad, Dr. E. Alshech, February 27, 2007.

25. Dr. Alshech writes: "A more recent indication of the increasingly organized nature of electronic jihad is an initiative launched January 3, 2007 on Islamist websites: mujahideen operating on the Internet (and in the media in general) were invited to sign a special pact called 'Hilf Al-Muhajirin' ('Pact of the Immigrants'). In it, they agree 'to stand united under the banner of the Muhajirun Brigades in order to promote [cyber-warfare],' and 'to pledge allegiance to the leader [of the Muhajirun Brigades].' They vow to 'obey [the leader] in [all tasks], pleasant or unpleasant, not to contest [his] leadership, to exert every conceivable effort in [waging] media jihad... [and to persist] in attacking those websites which do harm to Islam and to the Muslims...' This initiative clearly indicates that the Islamist

hackers no longer regard themselves as loosely connected individual activists, but as dedicated soldiers who are bound by a pact and committed to a joint ideological mission."

6   Mohajroon.com/vb/showthread.php?t=42442.

7   See MEMRI JTTM report Cyberspace as a Combat Zone: The Phenomenon of Electronic Jihad, February 27, 2007.

8   See MEMRI JTTM report Cyberspace as a Combat Zone: The Phenomenon of Electronic Jihad, February 27, 2007.

9   Al-Jinan.org.

0   Al-Jinan.org; Al-ekhlaas.net

1   Arabnews.com, October 14, 2001.

2   *The Open Society Paradox*, Dennis Bailey, 2005, p. 43; Arabnews.com, October 14, 2001.

3   "Islamic Cyberterror," Mark Hosenball, Newsweek, May 20, 2002, updated March 13, 2010; Strategicstudiesinstitute.army.mil/pubs/parameters/articles/03spring/thomas.pdf, Spring 2003.

4   "Cyberspace Full of Terror Targets," Tom Squitieri, USA Today, June 5, 2002; Strategicstudiesinstitute.army.mil/pubs/parameters/articles/03spring/ thomas.pdf, Spring 2003.

5   Part Two of the series was a landmark production that launched the Al-Qaeda strategy of encouraging individuals to carry out "lone-wolf" against targets in the West.

6   Special Dispatch No. 1326, Islamist Websites Monitor No. 9 – Mujahideen Gather Information on Anchorage International Airport, October 18, 2006.

7   Alnusra.net/vb/showthread.php?t=6946.

8   Alnusra.net/vb/showthread.php?t=21318.

9   Mohajroon.com/vb/showthread.php?t=38360.

0   *The Washington Post*, September 3, 2013.

1   MEMRI JTTM report In Issue V, Taliban Magazine 'Azan' Publishes Article On 'Counter-Drone Strategy', Seeks Jihadi Engineers Who Can 'Hack Into The Private Encrypted Network Of The Pentagon', March 28, 2014.

2   CNN.com, *Wall Street Journal*, December 17, 2009.

3   See MEMRI Special Dispatch No. 4448, Hizbullah Al-Manar TV Exclusive on U.S. Drone Captured by Iraqi Insurgents, January 25, 2012.

4   See MEMRI JTTM Member On Jihadi Forum: Lone Wolf Attackers Can Employ UAVs To Strike Sensitive Targets In U.S., West, October 27, 2014.

5   MEMRI TV clip no. 4570, ISIS Prisoner John Cantlie from Kobane: The Islamic State Has Nearly Taken the City, October 27, 2014.

6   See MEMRI JTTM report Online Jihadists Discuss Targets in U.S. - Including Government Buildings Like White House, CIA HQ, February 4, 2010.

7   See MEMRI JTTM report Jihadists Plan Cyber Attack on U.S. Government Computers, January 25, 2011.

8   See MEMRI JTTM report Jihadis Collaborate To Create "Center For Electronic Terrorism" Aimed At Virtually Attacking Infrastructure In U.S., U.K., France, June 13, 2011.

9   See MEMRI JTTM report Jihadi Confirms That Previously Established E-Jihad Brigades Project Is Growing And Developing, Alludes Brigades Participated In Recent Cyberattacks Against Israel, Planning Major Attacks Against Soft-Targets In U.S., April 16, 2013.

0   TCA Twitter account is @TN_cyberarmy

1   See MEMRI JTTM report 'Tunisian Cyber Army' And 'Al-Qaeda Electronic Army' Reportedly Hack Pentagon, State Department Websites, March 11. 2013.

2   Abu Dujana Al-Khurasani, aka Humam Al-Balawi, was the double agent recruited by Al-Qaeda who on December 30, 2009, blew himself up inside the CIA base in Khost, Afghanistan.

3   MEMRI JTTM report 'Irish' Jihadi Raises Possibility That Al-Qaeda Is Responsible For Alleged Cyberattack Against Open Source Center, April 2, 2013.

4   Shamikh1.info/vb/showthread.php?t=199519; Slashgear.com, April 13, 2013.

55   See MEMRI JTTM report Jihadi Launches Series Of Android Hacking Lessons, December 26, 2013.

56   Arabic.rt.com/news, May 21, 2013.

57   See MEMRI JTTM report In Response To U.S. Drone Attacks, Pro-IS Forum Posts Instructions For Disrupting, Downing Drones; Suggests Using Amazon Octocopter, September 4, 2014.

58   See MEMRI Inquiry & Analysis No. 704, Al-Qaeda's Embrace of Encryption Technology: 2007-2011, July 12, 2011.

59   CBS News, May 4, 2011.

60   See MEMRI JTTM report GIMF Announces Imminent Release of New Software, January 3, 2007.

61   See MEMRI JTTM report UPDATED: GIMF Releases Mobile Encryption Program For Secure Communication Between Jihadis, September 4, 2013 and MEMRI JTTM Report UPDATED GIMF Releases Encryption Software For Instant Messaging, February 07, 2013.

62   See MEMRI JTTM report GIMF Releases Updated Version Of Android Secure Communication App, Refers To The 'Cooperation Of Global [Communication] Companies With The International Intelligence Agencies' As Reason Behind Update, July 14, 2014.

63   See MEMRI Inquiry & Analysis No. 1086, Al-Qaeda's Embrace Of Encryption Technology - Part II: 2011-2014, April 25, 2014.

64   Abclocal.go.com, March 17, 2014; Huffington Post, March 18, 2014.

65   Justice.gov/usao/cae/news/docs/2014/2014_03/Teausant%20Complaint%20.pdf, accessed March 25, 2014.

66   See MEMRI JTTM report Jihadis Launch 'Military Electronics' Workshop, October 3, 2014.

67   See MEMRI JTTM report Jihadis Offer Tutorials On Obtaining Fake U.S. Phone Numbers To Maintain Access To Social Media, Circumvent Censorship, August 26, 2014.

68   See MEMRI JTTM report In Its New English-Language Magazine, Al-Qaeda Publishes Email And Public Encryption Key, Encourages Readers To Participate In The 'Effort To Revive The Spirit Of Jihad', October 21, 2014; see also MEMRI JTTM report Al-Qaeda in the Indian Subcontinent Launches English-Language Magazine, Calls For Targeting Western Economic Interests, October 20, 2014.

69   Ctc.usma.edu/wp-content/uploads/2012/05/SOCOM-2012-0000019-Trans.pdf.

70   See MEMRI JTTM report, "In Video Eulogy For Osama Bin Laden, Al-Qaeda Leader Ayman Al-Zawahiri Promises 9/11-Style Al-Qaeda Retaliation," June 6, 2011.

71   ABC, July 11, 2006 and June 2, 2008; *Washington Post*, June 23, 2006.

72   *Washington Post,* June 23, 2006.

73   See MEMRI JTTM report Senior Al-Qaeda Commander to Potential Recruits: Don't Come to Af-Pak, We Can't Afford to Train You, July 13, 2010.

74   It is noteworthy that the material presented in the document is extremely simplified and tailored for far less computer-savvy Internet users. TOR, on the other hand, is a greatly trusted tool amongst jihadis, mostly for its anonymizing qualities. The software has been covered in detail on jihadi forums. See MEMRI JTTM report AQIM Publishes First Installment In Electronic Jihad Series, September 20, 2013.

75   See MEMRI JTTM reports The 'Dark Web' And Jihad: A Preliminary Review Of Jihadis' Perspective On The Underside Of The World Wide Web, May 21, 2014; AQIM Publishes First Installment In Electronic Jihad Series, September 20, 2013; and Shumoukh Al-Islam Warns Its Members Not to Log In Without Taking Appropriate Cyber Safety Measures, December 7, 2011; also see MEMRI Inquiry & Analysis No. 1086, Al-Qaeda's Embrace Of Encryption Technology - Part II: 2011-2014, And The Impact Of Edward Snowden, April 25, 2014.

76   See MEMRI JTTM report **Technical Section Of Al-Battar Media Offers Tutorial On Installing TOR On Android,** October 27, 2014.

77   Youtube.com/watch?v=D8Mgpm1PgF4, accessed February 19, 2014.

78   Al-Kubi, Abu Asma, "Be a Mujahid." February 22, 2012. Accessed through Internet Archive, July 17, 2013; Ia601204.us.archive.org/27/items/Mujahid-Abu-Asmaa-Al_Kubi/Be_a_Mujahid.pdf.

79   See MEMRI JTTM report Al-Qaeda in the Arabian Peninsula Looks to Continue Political Assassinations, Encourages Lone Wolf Bombers in the West, November 5, 2009.

80   MEMRI Inquiry & Analysis No. 632, YouTube – The Internet's Primary and Rapidly Expanding Jihadi Base – Part IV, August 28, 2010.

81   MEMRI Inquiry & Analysis No. 576, Deleting Online Jihad and the Case of Anwar Al-Awlaki, December 30, 2009.

2 On November 9, Al-Awlaki published a post to his website calling Nidal Hasan a "hero." that same day, MEMRI released a special dispatch which provided details of Al-Awlaki's website, including the registration and domain, all hosted in the U.S., as well as contact information for these companies. Within hours, his website was shut down.

3 MEMRI Inquiry & Analysis No. 576, Deleting Online Jihad and the Case of Anwar Al-Awlaki, December 30, 2009

4 See MEMRI JTTM report U.S.-Born Iman Anwar Al-Awlaki: 44 Ways of Supporting Jihad, October 21, 2009,

5 MEMRI JTTM report Inspire 7 – American Jihadi Samir Khan: 'A Powerful Media Production is as Hard-Hitting as an Operation in America,' September 27, 2011.

6 See MEMRI JTTM report In Issue III Of English-Language Taliban Magazine 'Azan', German Militant Recounts Radicalization, Meeting With Al-Awlaki, Says: 'Brothers Are Working On Anti-Drone Technology' August 26, 2013.

7 See MEMRI JTTM report English-Language Jihadi Forum Publishes Eulogy for Anwar Al-Awlaki, Samir Khan; Online Jihadi Calls for 'Vengeance Raid' on American Websites, October 11, 2011.

8 See MEMRI JTTM report In Exclusive Interview In Inspire X, American Al-Qaeda Spokesman Adam Gadahn Tells 'Governments Of The Crusader West' To Withdraw From Islamic Lands; Says Arab Spring, Economic Crisis Signs Of Their Dwindling Power, Muslims' Ascendancy, March 1, 2013.

9 MEMRI JTTM, Indonesian Militant Umar Patek Urges Muslims To Go To Palestine To Wage Jihad, Calls For Jihad On Internet: 'This Is Not The Stone Age... This Is The Internet Era, There Is Facebook, Twitter And Others,' June 11, 2012.

10 See MEMRI JTTM report In Interview, Taliban Commander Reveals Details of Their Print Magazines, International Media Operations: 'We Are Also Active on Facebook and Twitter, Where We Publish News Every Day'; 'Wars Today Cannot Be Won Without the Media' – The Media is Directed At the Heart... [and] If the Heart Is Defeated, Then the Battle Is Won,' February 18, 2011.

11 See MEMRI JTTM report Taliban Spokesman Discusses The Taliban's Media Strategy: 'I Use Computers And Have Accounts On Facebook, Twitter, And YouTube; Winning over the Minds and Hearts of the Masses Who Visit The Websites' Is More Important, April 17, 2012.

12 See MEMRI JTTM Inquiry & Analysis Series, Report No.805, "Jihadi News Agency 'Kavkaz Center,' Affiliated With the Designated Terrorist Organization 'Caucasus Emirate,' Tweets Jihad and Martyrdom, January 24, 2012.

13 One example of this affiliation is an interview given by CE leader Umarov to Kavkaz Center. See MEMRI JTTM report Dokka Umarov: First We Will Retake All the Once-Muslim Regions, Then We Will Deal with Moscow, May 17, 2011.

14 State.gov/r/pa/prs/ps/2011/05/164312.htm.

15 State.gov/r/pa/prs/ps/2010/06/143579.htm.

16 MEMRI JTTM report In New Al-Qaeda Video, Captured Jewish American Aid Worker Warren Weinstein Urges Americans To Use Social Media To Secure His Freedom, December 27, 2013.

17 See MEMRI Inquiry & Analysis No. 724, Al-Qaeda, Jihadis Infest the San Francisco, California-Based 'Internet Archive' Library, August 17, 2011, and MEMRI Special Dispatch No. 4175, Following Al-Awlaki's Killing, His Legacy Lives On – Part I: In San-Francisco-Based 'Internet Archive' Library, October 3, 2011.

18 See the following major MEMRI research reports: MEMRI Inquiry & Analysis, Part I – Deleting Online Jihad and the Case of Anwar Al-Awlaki: Nearly Three Million Viewings of Al-Awlaki's YouTube Videos – Included Would-Be Christmas Airplane Bomber, Fort Hood Shooter, 7/7 London Bomber, and Would-Be Fort Dix Bombers; Part II – The Internet's Primary and Rapidly Expanding Jihadi Base: A Look at Al-Awlaki's Followers YouTube Pages; Part III – The Internet's Primary and Rapidly Expanding Jihadi Base: Taliban YouTube Page Remains Active; Part IV – The Internet's Primary and Rapidly Expanding Jihadi Base - Part IV: Young American YouTube Follower of Anwar Al-Awlaki on the Ground Zero Mosque and 9/11: 'America Reaps What It Sows'; 'You Pretend Like the World Trade Center and the Pentagon Was a Daycare Center or a Maternity Ward; If the People Who Did 9/11 Wanted To Kill Innocent People, They Would Have Bombed a School... Church... Daycare Center'; Part V: YouTube – The Internet's Primary and Rapidly Expanding Jihadi Base: One Year Later on YouTube – Anwar Al-Awlaki's Presence Expands, Al-Qaeda Goes Viral, Jihadists Post Thousands of Videos of Killing of U.S. Troops; European Jihadists Also Embrace YouTube.

19 See MEMRI JTTM report Jihadi Forum Offers Course in Uploading Materials to Youtube and Internet Archive – 'To Safeguard the Mujahideen's Legacy', August 1, 2011.

100 MEMRI Inquiry & Analysis No. 881, HASHTAG #Jihad Part II: Twitter Usage By Al-Qaeda And Online Jihadi Affiliated Groups Explodes; Apps Increasingly Used As Tools For Cyber Jihad, September 7, 2012; MEMRI Inquiry & Analysis No. 849, HASHTAG #Jihad: Charting Jihadi-Terrorist Organizations' Use Of Twitter, June 21, 2012.

101 MEMRI Inquiry and Analysis No. 881, HASHTAG #Jihad Part II: Twitter Usage By Al-Qaeda And Online Jihadi Affiliated Groups Explodes; Apps Increasingly Used As Tools For Cyber Jihad, September 7, 2012.

102 MEMRI Inquiry and Analysis No. 849, Hashtag #Jihad: Charting Jihadi Terrorist Organization's Use of Twitter, June 21, 2012.

103 See MEMRI JTTM report Syria-Based Saudi Sheikh Launches Second Campaign To Purchase Ammunition For Jihadi Groups Fighting In Syria, March 13, 2014, and MEMRI JTTM report Syria-Based Saudi Sheikh Launches Campaign To Purchase Ammunition For Jihadi Groups Fighting In Syria, October 29, 2013.

104 Facebook.com/alfork4n5.

105 Facebook.com/pages/%D9%85%D8%A4%D8%B3%D8%B3%D8%A9-%D8%A7%D9%84%D8%A3%D9%86%D8%AF%D9%84%D8%B3-%D8%A7%D9%84% D8%A3%D8%B9%D9%84%D8%A7%D9%85%D9%8A%D8%A9-%D8%AA%D9%82%D8%AF%D9%85/201050190062012.

106 Facebook.com/jihadul.ummmah?fref=ts.

107 This ring is said to have dispatched dozens of young mujahideen, some underage, from France, Belgium, and Morocco to fight in Syria. See MEMRI JTTM report An In-Depth Look At One Of The Facebook/Twitter Networks Of French-Speaking Jihadis Fighting In Syria: Glorifying Jihad In Syria, Condemning 'Heretical' France, February 17, 2014.

108 Publicintelligence.net/ufouoles-dhs-terrorist-use-of-social-networking-facebook-case-study/?utm_source=twitterfeed&utm_medium=twitter, December 5, 2010.

109 Publicintelligence.net/ufouoles-dhs-terrorist-use-of-social-networking-facebook-case-study/?utm_source=twitterfeed&utm_medium=twitter, December 5, 2010.

110 See MEMRI Jihad & Terrorism Studies Project Inquiry & Analysis Series Report No.650, Social Network Jihad Part 1, December 13, 2010.

111 MEMRI JTTM report Through Facebook Accounts And Internet Forums, Jihadi Groups Announce Q&A With Pakistan-Based German Jihadis, December 23 2013.

112 See MEMRI JTTM report AQAP Figure Recruited Kuwaitis, Saudis, And Others Via Facebook, Email And Over Phone, December 13, 2013 and MEMRI JTTM Report AQAP Confirms Sa'id Al-Shihri's Death, July 17, 2013.

113 See MEMRI JTTM report Member Of Al-Qaeda Affiliate Forum Announces First Jihadi Application On Facebook, March 11, 2013.

114 Reuters.com, January 22, 2014.

115 See MEMRI JTTM report Taliban Using Google Picture-Sharing Service Picasa To Publish Thousands Of Images Of Wounded U.S. Troops, Jihadi 'Martyrs', November 9, 2012.

116 MEMRI Inquiry & Analysis No. 816, Using Twitter, YouTube, Facebook and Other Internet Tools, Pakistani Terrorist Group Lashkar-e-Jhangvi Incites Violence against Shi'ite Muslims and Engenders Antisemitism, March 21, 2012.

117 See MEMRI JTTM reports A Terrorist's Use Of Social Media: American Al-Shabab Commander 'Omar Hammami (Abu Mansour Al-Amriki) On YouTube and Twitter, May 25, 2012; American Al-Shabab Commander 'Omar Hammami (Abu Mansour Al-Amriki) Publishes Part One Of Autobiographical Book, May 17, 2014; American Jihadi 'Omar Hammami Reportedly Killed By Al-Shabab Al-Mujahideen, September 12, 2013.

118 See MEMRI JTTM report AQIM Announces Launch Of 'Muslim Africa Blog', May 2, 2013.

119 See MEMRI JTTM report MOJWA Media Wing Launches New Jihadi Forum To Offer Training Courses In Bomb Manufacture, Booby-Trapping, January 14, 2013.

120 See MEMRI JTTM report ISIS Reportedly Testing Radio Broadcasts Of Religious Lessons, Group Statements, In Al-Raqqa, February 19, 2014.

121 See MEMRI JTTM report Jihadis Announce Establishment Of What They Call First Jihadi Internet Radio Station, January 8, 2013.

122 *Haq wa Ahsan Tafsir*, Ramadan 1435/June 2014.

123 See MEMRI Special Dispatch No. 5468, Danish Jihadi Fighting In Syria Tweets His Experiences In Jihad – Traveling To Europe And Back To Syria, Meeting Other European Foreign Fighters, October 3, 2013.

124 Ia902303.us.archive.org/14/items/shaaban1435_S_H_MM/shaaban1435_E.pdf, accessed September 10, 2014.

125 Dailymail.co.uk, February 18, 2014.

126 For more on the video game influence among foreign jihadis in Syria, see MEMRI Inquiry and Analysis No 1085, Modern Da'wa By Western Mujahideen On Social Media: Fighting In Syria Is 'Fun,' Thrilling, April 14, 2014.

127 See MEMRI JTTM report ISIS Declares Establishment Of Islamic Caliphate, Appoints ISIS Leader Abu Bakr Al-Baghdadi As 'Caliph': 'It Is Incumbent Upon All Muslims to Pledge Allegiance To The Caliph... And Support Him', June 29, 2014, and MEMRI JTTM report In New Message Following Being Declared A

'Caliph,' Islamic State Leader Abu Bakr Al-Baghdadi Promises Support To Oppressed Muslims Everywhere, Tells His Soldiers: 'You Will Conquer Rome [If You Follow My Advice], July 1, 2014.

[28] See MEMRI JTTM report Islamic State Publishes English-Language Magazine, Along The Lines Of AQAP's Inspire, July 5, 2014.

[29] See MEMRI JTTM report In Posthumous Appearance, Canadian Islamic State (IS) Fighter Urges Muslims In West To Travel To Syria, Says IS 'Can Easily Find Accommodation For You And Your Families,' July 13, 2014.

[30] See MEMRI JTTM report Kurdish Jihadi Group Ansar Al-Islam Announces Its Dismantling And Merger With Islamic State (IS), August 29, 2014.

[31] Mirror.co.uk, August 15, 2014.

[32] Mirror.co.uk, August 23, 2014.

[33] See MEMRI JTTM report Al-Baghdadi-Led Islamic State (IS) Tweets In Urdu For Audiences In Pakistan And India, July 15, 2014.

[34] Zamanalwasl.net, May 7, 2014.

[35] Youtube.com/channel/UC6jqoiHeKFABMjIz7QyhN8w, accessed May 9, 2014.

[36] See MEMRI JTTM report Islamist Syrian Opposition Group Establishes 'Electronic Islam Company' To Target Pro-Syrian Regime Webpages, May 9, 2014.

[37] See MEMRI JTTM report Tunisian Fighter And Writer For Jihadi Forum Died In Syria Fighting With Jabhat Al-Nusra; Previously, He Travelled Through West – And Then Used Google Earth To Join Jihad In Afghanistan, December 19, 2012.

[38] See MEMRI JTTM report Pakistani Province To Ban Skype, Viber, WhatsApp Over Terror Fears, Says Terrorists Have Switched Over From Regular Mobile Phone Services, October 4, 2013.

[39] *New York Times*, February 4, 2014.

[40] See MEMRI JTTM report An Inside Look At The Recruitment Of European Jihadis, March 25, 2014.

[41] See MEMRI JTTM report Islamist Opposition Battalion In Syria Provides Skype Contact Info To Potential Trainee Recruits, July 8, 2013.

[42] See MEMRI JTTM report Jihadis Move From Facebook To Yahoo's Flickr Picture Sharing Site – Uploading Images Of Martyrs, Including Foreign Fighters Killed In Syria From U.S., U.K., France, Australia, Spain, August 7, 2013; Islamist Opposition Battalion In Syria Provides Skype Contact Info To Potential Trainee Recruits, July 8, 2013.

[43] See MEMRI JTTM report New Fundraising Campaign To Provide Military Support For Jihadis In Syria, August 23, 2013.

[44] See MEMRI JTTM report Fundraising Campaign In Kuwait For Designated Terrorist Group Jabhat Al-Nusra, May 14, 2013.

[45] MEMRI JTTM Report: Pakistan-Based Jihadi Organization Lashkar-e-Taiba Using Exclusive Skype-Type Application To Plan Terror Attacks, December 4, 2013.

[46] See MEMRI JTTM report Senior Member of Shumoukh Suggests Using WhatsApp To Publish Jihadi Productions, Terrorize Allah's Enemies, September 3, 2013.

[47] See MEMRI Inquiry and Analysis No. 928, Hizbullah Circumvents August 2012 Bans By Apple And Google And Violates U.S., Dutch Law By Providing Ways To Download Its Apps For iPhone And iPad – Via Its Website – And For Android – Via The Netherlands-Based Aptoide And The TX-Based WhatsApp, February 08, 2013.

[48] CNN.com, September 17, 2014.

[49] WBFO.org, September 18, 2014.

[50] Investigativeproject.org/documents/case_docs/2482.pdf

[51] Investigativeproject.org/documents/case_docs/2482.pdf

[52] See MEMRI JTTM report Jihadi Group Embraces 'Google Plus' As Additional Means Of Disseminating Its Propaganda, July 20, 2012.

[53] See MEMRI Inquiry and Analysis No. 928, Hizbullah Circumvents August 2012 Bans By Apple And Google And Violates U.S., Dutch Law By Providing Ways To Download Its Apps For iPhone And iPad – Via Its Website – And For Android – Via The Netherlands-Based Aptoide And The TX-Based WhatsApp, February 8, 2013.

[54] See MEMRI Inquiry and Analysis No. 928, Hizbullah Circumvents August 2012 Bans By Apple And Google And Violates U.S., Dutch Law By Providing Ways

To Download Its Apps For iPhone And iPad – Via Its Website – And For Android – Via The Netherlands-Based Aptoide And The TX-Based WhatsApp, February 8, 2013.

155 See MEMRI JTTM report Taliban Using Google Picture-Sharing Service Picasa To Publish Thousands Of Images Of Wounded U.S. Troops, Jihadi 'Martyrs', November 9, 2012.

156 Telegraph.co.uk, December 9, 2008.

157 See MEMRI JTTM report Urdu-Language Jihadi Forums On Facebook, August 21, 2013.

158 See MEMRI JTTM report New Pro-Islamic State Of Iraq And Syria (ISIS) 'News' App For Android, Available At Google Play Store, April 22, 2014.

159 MEMRI Inquiry & Analysis No. 948, Online Jihadis Embrace Instagram, March 14, 2013.

160 MEMRI Inquiry & Analysis No.1065, Al-Qaeda And Its Offshoots Train And Indoctrinate Newborns, Toddlers, And Preschoolers – Captured On Twitter, February 7, 2014.

161 MEMRI Inquiry & Analysis No.1005, Jihadis Move From Facebook To Yahoo's Flickr Picture Sharing Site – Uploading Images Of Martyrs, Including Foreign Fighters Killed In Syria From U.S., U.K., France, Australia, Spain, August 6, 2013.

162 MEMRI Special Dispatch No. 5529, On Tumblr, British Member Of ISIS Who Recently Arrived In Syria Opens Q&A Session With Readers, Explains How He Reached His Destination, Quotes Bin Laden And Al-Awlaki, Speaks About His Jihadi Companions, November 20, 2013.

163 Mirror.co.uk, February 15, 2014.

164 See MEMRI JTTM report British Jihadi 'Abu Layth' Tweets About Going To Syria And His Experiences With Other Western Jihadi Fighters There, December 23, 2013.

165 See MEMRI Special Dispatch No. 5529, On Tumblr, British Member Of ISIS Who Recently Arrived In Syria Opens Q&A Session With Readers, Explains How He Reached His Destination, Quotes Bin Laden And Al-Awlaki, Speaks About His Jihadi Companions, November 20, 2013 and MEMRI JTTM report American ISIS Member From Chicago Active On Twitter And Facebook: 'Internet, Restaurants, Cars, iPhones... Allah Has Made... Jihad In Sham [Syria] So Easy', January 2, 2014.

166 Mirror.co.uk, February 15, 2014.

167 Mirror.co.uk, February 15, 2014.

168 See MEMRI JTTM report British Fighter In Somalia Promotes Jihad On Social Media, February 13, 2014.

169 See MEMRI JTTM report American ISIS Member From Chicago Active On Twitter And Facebook: 'Internet, Restaurants, Cars, iPhones... Allah Has Made... Jihad In Sham [Syria] So Easy', January 2, 2014.

170 Ask.fm/abumuhajir1.

171 See MEMRI Inquiry & Analysis No. 796, The Release of Top Al-Qaeda Military Strategist/Ideologue Abu Mus'ab Al-Suri from Syrian Prison – A Looming Threat, February 08, 2012.

172 *The Guardian,* August 21, 2014.

173 *USA Today*, July 17, 2013.

174 See MEMRI JTTM report Islamic State Releases Third Issue Of Its English-Language Magazine 'Dabiq,' Urges Muslims In The West To Emigrate To The 'Caliphate,' Taunts U.S. Over Foley's Death, August 30, 2014.

175 Charlierose.com/watch/60464014.

176 See MEMRI JTTM reports Islamic State (IS) Promotes Its Materials Via 'Media Points' Scattered Across Al-Raqqa, Syria, July 24, 2014 and Islamic State (IS) 'Media Points' Information Campaign In Ninawa, Which Uses Bluetooth And Other Technologies, August 7, 2014

177 Friendica.eu/profile/ale3tisam, accessed July 15, 2014.

178 Friendica.eu/profile/ajnad, accessed July 15, 2015.

179 Friendica.eu/profile/alhayaten, accessed July 15, 2014.

180 Friendica.com/

81. See MEMRI JTTM report Russian Jihadis Use Social Media To Raise Funds For Jihad In Syria, May 27, 2014.
82. See MEMRI JTTM report Hosted In Germany, Justpaste.it Is Being Widely Used By Terrorist Organizations To Publish Jihadist Content, September 3, 2014
83. Support.twitter.com/articles/227220-twitter-s-suggestions-for-who-to-follow#
84. Metronews.ca, March 19, 2013.
85. CNN.com, July 14, 2011.
86. Blogs.reuters.com, June 7, 2012.
87. Dailymail.co.uk, July 15, 2013.
88. See MEMRI JTTM report In Issue V, Taliban Magazine 'Azan' Publishes Article On 'Counter-Drone Strategy', Seeks Jihadi Engineers Who Can 'Hack Into The Private Encrypted Network Of The Pentagon', March 28, 2014.
89. Twitter.com/khorasan313, July 23, 2014.
90. See MEMRI JTTM report ISIS Declares Establishment Of Islamic Caliphate, Appoints ISIS Leader Abu Bakr Al-Baghdadi As 'Caliph': 'It Is Incumbent Upon All Muslims to Pledge Allegiance To The Caliph... And Support Him', June 29, 2014, and MEMRI JTTM report In New Message Following Being Declared A 'Caliph,' Islamic State Leader Abu Bakr Al-Baghdadi Promises Support To Oppressed Muslims Everywhere, Tells His Soldiers: 'You Will Conquer Rome [If You Follow My Advice], July 1, 2014.
91. See MEMRI JTTM report In Posthumous Appearance, Canadian Islamic State (IS) Fighter Urges Muslims In West To Travel To Syria, Says IS 'Can Easily Find Accommodation For You And Your Families,' July 13, 2014.
92. MEMRI JTTM report Islamic State (IS) Supporters Launch Social Media 'Blackout Campaign' To Protect IS Mujahideen, September 30, 2014.
93. See MEMRI JTTM report New English-Language Blog Suggests Using Bitcoin To Send 'Millions Of Dollars' To The Islamic State, July 7, 2014.
94. See MEMRI JTTM report The 'Dark Web' And Jihad: A Preliminary Review Of Jihadis' Perspective On The Underside Of The World Wide Web, May 21, 2014
95. See MEMRI JTTM report Jihadis Circulate Tutorials On Maintaining Online Anonymity And Security, Suggest Ways To 'Hide From The Crusader Alliance', October 28, 2014.
96. Youtube.com/watch?v=98FP4XgTydY, August 27, 2014, accessed September 18, 2014.
97. Hosted.ap.org/specials/interactives/wdc/documents/wh_terror060905.pdf, September 2006.
98. Whitehouse.gov/sites/default/files/counterterrorism_strategy.pdf, June 28, 2011.
99. Foreignpolicy.com, December 12, 2009.
100. See MEMRI Inquiry & Analysis No. 956, YouTube Questioned In U.K. House Of Commons Over Keeping Terrorism-Promoting Videos Active On Its Website; Of 125 Videos Of Al-Qaeda Commander Al-Zawahiri Flagged On YouTube By MEMRI, YouTube Keeps 57 Active, April 9, 2013.
101. Dailymail.co.uk, April 26, 2014.
102. France24.com, September 16, 2014.
103. BBC, October 7, 2014; AFP, October 9, 2014; Theregister.com, October 10. 2014.

ABOUT THE MEMRI CYBER & JIHAD LAB (CJL)

RESEARCH
The CJL monitors, tracks, translates, and researches cyber jihad originating from the Middle East, Iran, South Asia, and North and West Africa

ANALYSIS
The CJL translates information from Arabic, Farsi, Urdu, Pashtu, Dari, and other languages into English and produces detailed analyses

SOLUTIONS
The CJL innovates and experiments with possible solutions for stopping cyber jihad

LAB.OR.A.TO.RY
"A place equipped for experimental study in a science or for testing and analysis"
— MERRIAM-WEBSTER

TRANSLATING:
ARABIC   URDU   DARI
FARSI   PASHTU

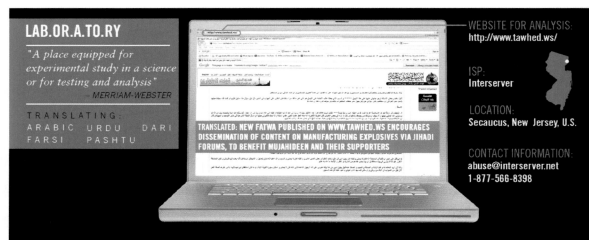

TRANSLATED: NEW FATWA PUBLISHED ON WWW.TAWHED.WS ENCOURAGES DISSEMINATION OF CONTENT ON MANUFACTURING EXPLOSIVES VIA JIHADI FORUMS, TO BENEFIT MUJAHIDEEN AND THEIR SUPPORTERS

WEBSITE FOR ANALYSIS: http://www.tawhed.ws/

ISP: Interserver

LOCATION: Secaucus, New Jersey, U.S.

CONTACT INFORMATION:
abuse@interserver.net
1-877-566-8398

TRANSLATION: Minbar Al-Tawhid wa'l-Jihad ("Pulpit of Monotheism and Jihad")

CONTENT ANALYSIS: Pro-jihad website providing religious opinions, including justification for Western Muslims seeking to carry out lone-wolf attacks

WEBSITE OWNER: Abu Muhammad Al-Maqdisi, a Jordanian currently imprisoned in Jordan, is regarded as the most influential Salafi-jihadi scholar today; he was spiritual mentor to the late Al-Qaeda in Iraq leader Abu Mus'ab Al-Zarqawi

MEMRI IS THE ONLY ORGANIZATION DOING THIS VITAL WORK

WHAT DOES THE CJL ANALYZE?

JIHAD/TERRORIST GROUPS ON U.S. SOCIAL MEDIA

HACKING GROUPS FROM MIDDLE EAST AND SOUTH ASIA

IRAN CYBER INITIATIVE
Farsi-language Iranian government cyber initiatives against the West.

FATWAS ON HACKING
"If you want to destroy it, and have the ability to do so, it's okay ... because it is an evil website."
— ABD AL-AZIZ AL-SHEIKH, SAUDI GRAND MUFTI SHEIKH

JIHAD/TERRORIST SITES & ISPS
Jihadi Tech and Military Monitor analyzes first-tier Al-Qaeda military- and technology-related message boards.

202-955-9070 • MEMRI@MEMRI.ORG
WWW.MEMRI.ORG/DONATE
P.O. BOX 27837 | WASHINGTON, DC 20038
MEMRI is a 501(c)3 organization. Your donation is tax deductible.

HOW THE CJL CRAFTS SOLUTIONS FOR COMBATING CYBER JIHAD

 INITIATIVES
Companies and individuals concerned about cyber jihad researching and innovating projects in their areas of expertise

 LEGISLATURE
Advancing legislation and initiatives federally and on the state level – including Capitol Hill and attorneys-general – to draft and enforce measures that will serve as precedent for further action

 TASK FORCE
Bringing together and working with leaders in business, law enforcement, academia, and families of terror victims

 TECHNOLOGY
Recruiting and working with technology industry leaders to craft and support efforts and solutions to combat cyber jihad

 EDUCATION
Working to increase awareness and cultivate in-depth understanding of the problem so that appropriate solutions can be crafted

 BUILDING A COMMUNITY
Fostering involvement among all those concerned about the issue and enlisting their help in a global effort to inform politicians, media, academia, the business community, and others on the issue

www.memri.org/CJL

Made in the USA
Columbia, SC
08 September 2020